H_2O
and the
Waters of Forgetfulness

By the same author

Celebration of Awareness
Deschooling Society
Tools for Conviviality
Energy and Equity
Limits to Medicine
Disabling Professions
Shadow Work
Gender
The Right to Useful Unemployment
ABC: The Alphabetization of the Popular Mind (with Barry Sanders)
In the Mirror of the Past,
Lectures and Addresses 1978-1999

Ivan Illich

H_2O
and the
Waters of Forgetfulness

MARION BOYARS
LONDON • NEW YORK

Reprinted in 2012 by
Marion Boyars Publishers
26 Parke Road
London SW13 9NG

First published by Marion Boyars in the United Kingdom in 1986. This text was first published by the Dallas Institute of Humanities and Culture, Dallas, USA.

© Ivan Illich 1986

All rights reserved.
Available in the United States of America.

The right of Ivan Illich to be identified as the author of this work has been asserted by him in accordance with the Copyright, Designs and Patents Act 1988.

No part of this publication may be reproduced, stored in a retrieval system, or transmitted, in any form or by any means, electronic, mechanical, photocopying, recording or otherwise, except brief extracts for the purposes of review, without the prior written permission of the copyright owner and publisher.

Any paperback edition of this book whether published simultaneously with, or subsequent to the hardback edition, is sold subject to the condition that it shall not, by way of trade, be lent, resold, hired out, or otherwise be disposed of, without the publishers consent, in any form of binding or cover other than that in which it is published.

British Library Cataloguing in Publication Data is available from the British Library.

ISBN: 978-0-7145-2854-0

About the Series

Ideas in Progress is a commercially published series of working papers dealing with alternatives to industrial society. Authors are invited to submit short monographs of work in progress of interest not only to their colleagues but also to the general public. The series fosters direct contact between the author and the reader. It provides the author with the opportunity to give wide circulation to his draft while he is still developing an idea. It offers the reader an opportunity to participate critically in shaping his idea before it has taken on a definite form.

Future editions of a paper may include the author's revisions and critical reactions from the public. Readers are invited to write directly to the author of the present volume at the following address:

        Ivan Illich
        Apdo. 479
        Cuernavaca, Mor.
        Mexico

By the same author

Celebration of Awareness
Deschooling Society
Tools for Conviviality
Energy and Equity
Limits to Medicine: Medical Nemesis – The Expropriation of Health
Disabling Professions
The Right to Useful Unemployment
Shadow Work
Gender

Acknowledgments

In May 1984 Gail Thomas and her colleagues invited me to Dallas for a short lecture. I promised to correct my manuscript for publication, and it turned into this book. This book, however, is not based directly on the lecture but rather on a German version of it that Ruth Kriss-Rettenbeck has prepared. Only in another language and through her questions did I come to realize the potential that lay in my original text. I am grateful to The Dallas Institute of Humanities and Culture for printing the galleys of this book in time for a seminar on the "Historical Heuristics of Body Images" that I was then conducting with Barbara Duden. Susan Hunt has revised every line of my text and Beverly Hall gave shelter and assistance while I corrected the final manuscript. I admire Scott and Susan Dupree for their ability to typeset a book from my manuscript that resembled a quilt.

What the reader will find in these pages is literally an "essay", an attempt: a text surrounded by glosses, tangents, and marginalia that I have added in the course of many conversations.

Dallas Town Lake

I understand that in Dallas during the last seventy years numerous citizen groups have urged the construction of a midcity lake. At the beginning of the century such a project would have amounted to a modest enterprise; now it has taken on the dimensions of an extravaganza. During 1984, yet another group of experts has been mulling over the technical feasibility and social acceptability of drowning a dozen midtown blocks. Proponents of the lake anticipate that it will irrigate business and fantasy, taxes and recreation; opponents consider the proposal an elitist misappropriation of public funds. Among the many arguments that have been raised, tabled, and warmed over for seven decades, one stands out. Both those who want to push and those who want to stop the lake imply that the natural beauty of a body of water would be morally uplifting to the civic life of Dallas.

The Nude in the Tub

The popular wisdom which holds that water possesses "natural beauty" and that this beauty has impact on civic morale is not always overtly expressed. However, you have only to poke fun at the belief in the civic magic of a body of water, and people react as if you had made a dirty joke. This, I claim, is so because water, which has always been perceived as the feminine element of nature, in the nineteenth century was tied to a new "hygienic" image of woman, which was itself a creation of the Victorian age. Only the late nineteenth century tied female nudity as a cultural symbol to the tap water of the bathroom. The proximity of suds and nude in the bath domesticated both water and flesh. Water became that stuff that circulates through indoor plumbing, and the nude became the symbol of a new fantasy of sexual intimacy defined by the newly created domestic sphere.

Ivan Illich

The evolution of the subtle ties between water and the nude can be observed, in all its complexity, in the paintings of the period. The painter found it less and less necessary to justify the nude by presenting her in religious or mythological terms. By showing her as bather he could merge woman and water as part of "nature." Only a rare genius such as Courbet could successfully paint *The Source* as a woman of incredible specificity, utterly lacking in self-consciousness yet bluntly assertive of her flesh. For the run-of-the-mill painter, this association of flesh with water served to render the feminine body innocuous. First, in the course of Ingres's long life, the term *nude* became synonymous with the Turkish bath. Then the aging Degas filled his atelier with tubs, bowls, and basins in which to pose his models. His pastels offer a historical source for the domestic bath during the late nineteenth century. It is not so much the nude he paints as woman's absorption in the relationship of her body to the water with which she sponges herself.[1]

1. Franco **Russoli** lists the pastel paintings of nudes executed from 1878 to 1890 as numbers 867 through 957 and 990 through 1054. Huysmans accused Degas of a disdainful hatred for the female body when a number of these pastels were shown at the last Impressionist exposition in 1886. As with his earlier dancer subjects, these women who bathe, wash, towel, or groom themselves are shown at the moment when a difficult movement has reached an unstable balance.

A new relationship develops in Degas between the nude and its background. While Rembrandt places his "Susannah" in a landscape which is continuous, both spatially and emotionally, with the feelings of a startled girl, and Manet, deliberately subverting this model, seats his surprised nymph in front of a studio background that is discontinuous with her (see **Krauss 1967**), in Degas tub and towel, soap and water as they touch her skin are at the center of the bather's attention. Degas keeps objects to a necessary minimum in order to indicate that this contact between the nude and water happens within the domestic sphere. The skin depicted by his strokes mimics the surface quality of water. "The women I paint are simple, decent persons, totally absorbed in the care of their bodies," he noted in his diary. His bathers are as absorbed in the care of their nudity as Degas seems to be in his own technique.

H_2O and the Waters of Forgetfulness

The intertwining of urban water and the nude constitutes one of the strands of a taboo woven to protect the symbolism of public water use from analysis. We may, for instance, debate quite openly our selection of the architect who will dress up the stuff that runs through Dallas pipes. We feel free to criticize the way he displays it, makes it dance or sparkle. But we do not feel free to question the natural beauty of water itself because we know, yet cannot bear to acknowledge, that this "stuff" is recycled toilet flush.

The Historicity of "Stuff"

I want to question the beauty intrinsic to H_2O because The Dallas Institute of Humanities and Culture has offered to make its own telling contribution to the dispute over Town Lake. We have been invited to discuss "water and dreams" insofar as they contribute to "making the city work." The title of this conference was taken from a book just translated and published by members of the Institute. *Water and Dreams* was written forty years ago by Gaston Bachelard.[2] It is one of a series of essays in which he analyzes the way we imagine matter, that "stuff" to which our imagination gives shape and form. I shall continue along the lines of Bachelard's investigation, distinguishing "stuff" and its form, and reflect on the bond the imagination creates between two kinds of stuff from which a city is made: urban space and urban water.

The interrelationship between water and space may be explored on two different levels. The first deals with *form*. On

---

According to Eldon N. **Van Liere**, "The image of the bather provides the story within a story of French painting in the nineteenth century," a drama interior to the history of art but consonant with the main themes of that history. On Courbet consult **Farwell**.

2. Gaston **Bachelard**, *Water and Dreams*. I follow the translator of **Husserl** (295ff.) in choosing the English term "stuff," even though I do not address the same epistemological sense treated there.

this level comparison focuses on the common aesthetic features a period's imagination has given to urban water and to urban space. An epoch's contribution to the style of their perception and representation is at the core of this approach to poetry or painting, sculpture or dreams. The question is "How did baroque art use or show water?" not "What does the epoch believe water *is*?" Water itself, on this first level, has no history; since "the beginning, when the earth was unsightly and unfinished," water was H_2O. According to this hypothesis all stories of creation from around the world tell about the origin of the same stuff, since the "stuff" as such is a-historical.

I do not intend to explore water in this fashion—nor, for that matter, space or the imagined bond that unites them.[3] From the start I shall refuse to assume that all waters may be reduced to H_2O. I will not deal with city space as though it could be universally defined in terms of Cartesian coordinates or of census criteria. For not only does the way an epoch treats water and space have a history: the very substances that are shaped by the imagination—and thereby given explicit meanings—are themselves social creations to some degree.

I want to explore the historicity of matter, the sense that an epoch's imagination has given to the canvas on which it paints its imaginings, to the silence of a room into which it projects its music, to the space that it fills with the aura that

3. The ideas that (1) waters, airs, and places are conceived by the creator for man and that (2) they correlate with each culture and codetermine its historical uniqueness have, according to **Glacken**, dominated all preindustrial societies. By the late seventeenth century man begins to be understood as a geographic agent, and only during the later twentieth century does the social perception of the landscape by a given epoch come to be understood as a social force that does not just reflect society's style but reinforces and shapes its sense of reality. The argument is subtly and strongly made by **Fabricant** and has been recently illustrated vividly by **Stafford** and also **Jordanova**.

it can taste or smell.[4] The attempt to do so is not new: and the evidence that it always fails is no reason for refraining from trying again, to write the history of life's widow as Luis de Sandoval y Zapata, a seventeenth-century Mexican, calls this "stuff" in a baroque poem translated by Samuel Beckett.

To Primal Matter

Within how many metamorphoses,
matter informed with life, hast thou had being?
Sweet-smelling snow of jessamine thou wast,
and in the pallid ashes didst endure.

Such horror by thee to thyself laid bare,
king of flowers, the purple thou didst don.
In such throng of dead forms thou didst not die,
thy deathbound being by thee immortalized.

For thou dost never wake to reason's light,
nor ever die before the invisible
murderous onset of the winged hours.

What, with so many deaths art thou not wise?
What art thou, incorruptible nature, thou
who hast been widowed thus of so much life?

Water as "Stuff"

The substance that is considered "water" or "fire" varies with culture and epoch. And water is always dual. It tends to stand for the original couple—more often than not for the twins who before creation lay in each other's arms. Water envelops what exists before space was. Water is the blood that nourishes even before milk can flow. Many things can be waters: there are some cultures in which the salty ocean is as unlike blood as it is unlike the water that quenches thirst. And there are jungle cultures in which heaven and earth are

4. Andre **Malraux**, in *La Tentation de l'Occident*, speaks in this vein:
> Vous avez distingué dans l'homme certains sentiments, et leurs causes les plus communes; mais vous voyez qu'il y a, dans ce que vous appelez *homme* quelque chose de permanent qui n'existe pas. Vous êtes semblables à des savants fort serieux qui noteraient avec soin les mouvements des poissons, mais qui n'auraient pas découvert que ces poissons vivent dans l'eau.

perceived as just so many different manifestations of water. Among the Indians on the Venezuelan border of Brazil, even the dead turn to water after thrice seven years to return to earth as women, who are perceived as dew.

Even the border between water and fire can shift. In Vedic mythology *soma* is the fire that envelops all being and that flows and ebbs around the sun; it is fire that can be drunk. In Arabic *al Ko'hol* is a fine metallic powder that is sublimated from mercury and used to embellish women; when applied as a shade to the eyelids it renders them intoxicating. Only after Paracelsus had distilled *alcohol* from wine was its intoxicating power ascribed to a spirit of water. Thus the very "stuff" that is watery, no less than its form, lies in the eyes of the beholder.

In making this distinction between imagination as the source of form and imagination as the wellspring of formless "stuff" I am building on a foundation established by Gaston Bachelard. In his writings he returns again and again to a fundamental contrast between two mutually constitutive aspects of imagination: a formal one and a material one. The form and matter of our imagining cannot be understood separately because one cannot exist without the other. But the fact that we cannot separate our experience of passion from the element of fire and cannot imagine fire without passion in no way implies that the two are at all times perceived as versions of the same principle. Love, the hearth, rage, war, and passion are kindled. They are set aflame by contact with a "stuff" that is imagined as fire. In each culture the line that separates the inflammable from the fireproof divides reality in a different way. In the south of Mexico there are two tribes which share the same territory: in one tribe women inflame men's desire and in the other they liquefy men's innards. But in both beneath the mass of images, verbal variations, moods, tactile experiences, and lights that shape water in our imagination, there is a stable, dense, slow, and fertile watery stuff that obscurely vegetates within us. It lies beyond the reach of any one of our separate senses: "its black flowers bloom in matter's darkness" and become visible when the imagination lets them "sing reality." The time has come for historians to begin listening to "the sonority of these dormant waters" (Bachelard) to become sensitive to the history of matter.

Following dream waters upstream, the historian will learn to distinguish the vast register of their voices. As his ear is attuned to the music of deep waters, he will hear a discordant sound that is foreign to waters, that reverberates through the plumbing of modern cities. He will recognize that the H_2O which gurgles through Dallas plumbing is not water, but a stuff which industrial society creates. He will realize that the twentieth century has transmogrified water into a fluid with which archetypal waters connot be mixed. With enough money and broad powers to condemn and evict, a group of architects could very well create out of this sewage a liquid monument that would meet their own aesthetic standards. But since archetypal waters are as antagonistic to this new "stuff" as they are to oil, I fear that contact with such liquid monumentality might make the souls of Dallas's children impermeable to the water of dreams. In voicing this fear, I am not arguing against the construction of a lake that would provide moorings for inexpensive rowboats, cool the city, and sparkle at night. Pleasure boats, temperatures, and the reflection of skyscrapers are not my concern here. I want to deal with waters and dreams. I want to explore the moral and psychological consequences that will flow from the public display of recirculated toilet flush with pretense to the aesthetic symbol of a wedding between water and urban space.

Ivan Illich

Dwelling Space: Neither Nest nor Garage

The same distinctions concerning the smell, sight, taste, and tactility of this ineffable stuff called water can also be applied to urban space. Each culture shapes its own space, the very space it engenders in becoming a culture. Space is not, as Durkheim says in one brilliant passage, the homogeneous environment that the philosophers have imagined.[5] Space is a social creation which results from the all-embracing asymmetrical complementarity enshrined in each culture.

"Where do you *live?*" and "Where do you *dwell?*" are synonymous. They remain so in most translations into other, even nonwestern, languages. This unusual constancy of meaning indicates that "living" and "dwelling" have traditionally implied one another; one stresses the temporal, the other the spatial aspect of being. To dwell means to inhabit the traces left by one's own living, by which one always retraces the lives of one's ancestors. "Dwelling" in this strong sense cannot really be distinguished from living. From day to day dwellers shape the environment. In every step and movement people dwell. Traditional dwellings are never terminated. Houses constantly grow; only temples and palaces can

---

5. "Space is not that vague and indeterminate milieu that Kant had imagined: in this pure and absolutely homogeneous form it would be totally useless, and not even thought could grasp it. Spatial representation consists essentially in a first coordination of sensible experience. However, such a coordination would not be possible if the regions of space were qualitatively equivalent, if they were really such that one could be substituted for the other. In order to place things in space, it is essential that one place them differently, some to the right and some to the left . . . some up there, others down there. . . . Space could not be what it is if it were not divided and differentiated. . . and these differences seem to come from the fact that a different emotional value is assigned to these regions. And since all people who belong to the same civilization imagine space in the same way . . . it is unavoidable that their emotional values should also be alike, that they be—almost inevitably—of social origin" (Emile **Durkheim**, 15).

H_2O and the Waters of Forgetfulness

be "finished." Dwelling means living insofar as each moment shapes a community's own kind of space.[6]

The sort of dwelling that results from this vernacular ac-

6. The opposition between vernacular dwelling space and industrial building sites is obvious. The first is a vernacular "commons" regulated by custom, the second an economic commodity that can be regulated only by formal laws. This obvious distinction was turned into a public argument through a series of papers published by John **Turner**. On several occasions I have tried to add an historical perspective to the practical insights with which John Turner has influenced thinking among planners and architects.

Dwelling-space has a number of characteristics, each of which allows a contrast with that housing space in which people are merely stored. Dwelling-space is (1) confining, (2) concentric, (3) gendered, and (4) governed by custom.

1. **Chaunu** has studied rural space in France from the thirteenth to the eighteenth centuries and has found that ninety percent of the material existence of each village was the result of activities going on within a radius of about two miles. This radius was characteristic of preindustrial Europe. In comparison with other peasant cultures, it is rather large. The Indian *desa* (village) has a radius measured by the distance that a bullock can comfortably reach for plowing; this is one reason why the cow is its sacred symbol. Its surface is about one-third of the post-medieval European standard that fits the measure of donkey and horse.

2. **Karnoch** has shown that as late as 1950 French villagers perceived space as constituted by three circles of social distance: the village space, in which most life goes on; the "valley," whose inhabitants are treated as outsiders but not as strangers, and the region in which only individual families are related by kinship to the village.

3. Dwelling space is always a "gendered" milieu, the result of a dyssymmetric complementarity between a man's and a woman's domain. On this subject see **Illich**, *Gender*, 105ff. and **Rogers (1979)** and **Ardener (1981)**. For an introduction to dyssymmetry in physics, see **Frisch**.

4. Each layer of vernacular space—rural or urban—is generated by a distinct set of customs and made visible by distinct rituals. These customs and rituals establish the experience of closeness or distance between family members, neighbors and villages, and the "space" which results from this experience is non-Cartesian. For examples see A. A. **Ott** and **Jolas**. Language strongly reflects the vernacular subjectivity of dwelling space: **Verdier**. People from different cultures cohabiting in the same place can live in distinctly perceived dwelling spaces. See, e.g., **Greverus, Boughali, Pingaud, Mounin, Lawrence**.

tivity must be carefully distinguished from an animal's lair and—just as much—from a merchant's storage. Animals are born with the instinct which dictates their behavior. The nest or the web, the den or the hole are created by the animal in the harness of its genes. Dwellings are not such lairs for breeding; they are shaped by a culture. No other art expresses as fully as dwelling that aspect of human existence that is historical and cannot be reduced to biological programs. But just as a dwelling is not a special spot determined by a territorial instinct, it is also not a garage.

This second point is as important as the first because active dwelling has become nearly impossible in Dallas. Dallas's citizens have lost the potency to imprint their lives on urban space. They use or consume their "housing." One must be quite wealthy to be able to relocate a wall in one's house. We need not necessarily deplore this circumstance, but we must be willing to explore it. Most people today do not dwell in the place where they spend their days and leave no traces in the place where they spend their nights. They spend their days next to a telephone in an office and their nights garaged next to their cars. Even if they wanted to dwell in the traditional manner, the material from which Dallas is made would not register their traces. The traces people manage to leave in the course of living are perceived as dirt that must be removed, as wear and tear that calls for repair, as the devaluation of a considerable investment. Dallas's space is not only "safe," innocuous for the transient, it is "man-proofed": it is hardened against defacement by contact with life. The census tracts that constitute Dallas do not, for this reason, make up a dwelling space. Children grow up and die without ever having had a chance to experience living-as-dwelling. The ability to dwell is a privilege of the dropout.

Only in so-called developing countries is dwelling still within the reach of ordinary people. Some of the poor who inhabit impoverished nations may still dispose of dwelling space. From the perspective of the new bureaucratic and pedagogical keepers, their inhabitants still "live on their own dirt." From the air you see the anarchic patches, on which life still shapes space. But inexorably development turns shanties into a slum. The recognition that a new gulf between "living" and "dwelling" has made them into separate activities that have been both previously unknown can lead to self-pity—but also to action. It leads some to romantic hankering after a lost "wholeness"—I want to make it into the starting point from which I explore the conditions that might allow a partial reconstitution of urban dwelling space.

Here I am focusing on water in order to reflect on one such condition. It is not water as a commodity that is at issue, nor its waste, its pollution, the ecological consequences of its irresponsible extraction, the biological consequences of poisoning it, or even its maldistribution—which means that, in Mexico City, sixty percent of all water is given to three percent of the households, and fifty percent of the households make do on five percent. These are also crucial issues, but they deal with water in a different sense. The water I speak of is the water needed for dreaming city as a dwelling place.

I shall first enlarge on the nature of living space. Then I shall comment on three typical kinds of urban space, relating each to a different kind of water: well, piped, and circulating water. In the end I shall return to the original question concerning the recuperation of dream water by the city child.

The Ritual Creation of Space

The imagination is not—as its etymology might suggest—the faculty of forming one's images of reality. It is, rather, the faculty of forming images of the invisible; it is the faculty that "sings reality." The classical town is first and foremost a ritual song of this sort. Its wellspring is dreams. Every urban culture seems to have its own ritual proceedings through which this dream of "life as an indwelling flow" is reflected in the social representation of in-habitable space. An agglomeration of huts or tents turns into a settlement or town only when its space has been recognized ceremonially as substantially other than rural expanse, when it is opposed to the "outside," when the paths that transverse its space are recognized as roads. For anyone who wants to understand the formation of inner space and urban form in Western culture from an anthropological perspective, the most prudent and learned guide is Joseph Rykwert, especially his definitive description in *The Idea of a Town*. I draw on his insights into the rituals which, in the classical world in general and in Etruria in particular, have created urban space.[7]

In the classical tradition, the founding of a town begins with the calling of its founder, usually in a dream. The culture hero Heracles appears to Myskelos in a dream and appoints him to found a colony, quite against the will of his neighbors and the laws of his Achaian town. When his project is brought to a vote, the god must go so far as to cheat and exchange the black "no" stones for white "yes" ones. Most founders are led by a sign in a dreamlike state to the site where the new town will be. Sometimes a wounded game animal, a strange bird, a cloud, or lightning takes him to the

---

7. On the detailed description of inaugurations, **Rykwert** is reliable. Sources are more amply quoted in **Nissen** and in **Wolfson** (v. 1, 242ff.). **Norden** deals with the rituals related to houses. The idea of inauguration has had a very important afterlife in Christian theology and allegory. It is reflected in the liturgy used in the consecration of medieval churches; see **Ohly**, 255ff. **Maurmann** deals mainly with the orientation of the celestial Jerusalem in Hildegard von Bingen and Honorius Augustodunensis; see also **Novotny**. In inaugural terminology, the "tescum" was opposed to the "templum." The full sense of "tescum" remains uncertain. Varro, L.G. 7, 10, gives "loca quidam aperta, quae alicuius dei sunt" (open spaces that belong to some god). On mystical space, see also **Landolt** and **Corbin**.

H_2O and the Waters of Forgetfulness

spot chosen by the gods. Aeneas follows the sow to the place where she drops her litter and where Alba Longa will stand. In a dreamy utterance the Pythia foresees the destiny of a settlement. She sends Myskelos to Kroton to make space for Pythagoras and destines the merchant Archias to become wealthy in Syracuse and die there at the hand of his lover. The dream of foundation is always pregnant with destiny, though only obscurely.

However, neither the vocation of a founder nor a mandate from the oracle at Delphi nor even the actual settlement of a site suffices to make a locality into a town. The intervention of a recognized seer is required, an augur who creates space at the site discovered by the founder. This social creation of space is called in-auguration. The augur is specially gifted: he can see heavenly bodies that are invisible to ordinary mortals. He sees the city's templum in the sky. This term is part of the technical vocabulary of his trade. The *templum* is a polygonal shape that hovers over the site found by the founder and that is visible only to the augur as he celebrates the inauguration. The flight of birds, a trail of clouds, the liver of a sacrificed animal can assist the augur in the *con-templatio*, the act in which he projects the figure seen in the sky onto the landscape chosen by the god. In this *con-templatio* the heavenly *templum* takes its this-worldly outline.

But *con-templatio* is not enough. The outline of the *templum* cannot settle upon the earth unless it is properly con-*sidered*, aligned with the stars (*sidus*). Con-*sideratio* follows *con-*templatio. *Con-sideratio* aligns the *cardo* (the axes) of the *templum* with the city's "star." The *cardo* was originally a "hinge" with an explicit, concrete, masculine–feminine symbolism.

The in-auguration is concluded by the naming of those parts of the city that will be right and left, front and back, and by providing a content for the spaces thus envisioned, fixing (*de-signatio*) the place for a *mundus*, or mouth of the underworld, which opens near the *focus*, the focal (fire) gate to the other world, where the Erinyes can surface. However, none of the augur's gestures or signs leave any visible trace on the ground. They are fixed in models of livers or of wheels, some of which have come down to us. The augur's actions constitute an incantation of space by the opposition and wedding of right and left which has yet to be made tangible. The founder himself must perform the wedding between this dyssymmetric *templum* and the landscape.

For this ceremony two white oxen are hitched to a bronze plow, the cow on the inside, drawing the plow counterclockwise, thus engraving the *templum* into the soil. The furrow creates a sacred circle. Like the walls that will rise on it, it is under the protection of the gods. Crossing this furrow is a sacrilege. To keep this circle open, the plowman lifts the plow when he reaches the spots where the city gates will be. He carries (*portat*) the plow to create a *porta*, a doorway. Unlike the furrow and walls guarded by the immortals, the threshold and gate will be under civil law. At the *porta*, *domi* (dwelling space) and *foras* (whatever lies beyond the threshold) meet; the door can swing open or be closed. Benveniste remarks that there is a profound asymmetry between the two terms in Indo-Germanic languages;[8] they belong to unrelated sets of words. They are so distant from one another in meaning that they cannot even be called anti-

8. It would be a mistake to assume that the rituals which create interior space necessarily inaugurate a building. The following example from pre-Hindu India will illustrate this point.
Kramrich (1968) describes the making of Indian floor paintings (*dhuli chitra*) executed on the ground with rice paste. Their making is an exclusive privilege of women, and the skill is handed down from mother to daughter, whose training begins in the fifth or sixth year. Before she can marry at twelve, a girl must have reached full competence. These "yantras" have not changed much since pre-Hindu times and are native to old India. They are still widely practiced by Brahmin women. Each yantra

thetical. *Domi* refers to in-dwelling, while *foras* refers to whatever else is shut out.

Only when the founder has plowed the *sulcus primigenitus* (furrow) around the future town perimeter does its interior become space that can be trodden and only then is the arcane celestial *templum* rooted in the landscape. The drawing of the *sulcus* is in many ways similar to a wedding. The furrow is symbolic of a hierogamy, of a sacred marriage of heaven and earth. The *sulcus primigenitus* carries this meaning in a special way. By plowing a furrow around the future town, the founder makes inner space tangible, excludes outer space by setting a limit to it, and weds the two spaces where the walls will rise later.[9]

---

forms a will directed to an end: within the yantra an invoked invisible presence finds its allotted place. A woman's purpose is confined and controlled and isolated from the ground. In the magical circle and sacred squares of the yantra, power is spellbound, cannot escape, and thus creates space. These yantras do not form abstract patterns; they are the shape of conceptions. They are intuited functional diagrams transmitted by women. The moon, the sun, the stars, and earth are integrated in them, along with the things desired by the young woman. The whole cosmos is conjured up to bless and fulfill them. Although the yantras were not sanctioned by the Vedas, they are customarily considered essential to temple building. But today when such pre-Aryan mandalas are made by male tantriks, those who make them have to abstain from all food prepared by woman; they must not so much as hear the sound of bangles. Kramrich asks, "Is it partly because these men have to guard their art from those who had the power to evolve it?"

9. **Benveniste** attempts through etymological studies to recover the social meanings carried by vocabulary: "We are throwing some light on meanings; others will be concerned with what words designate"(I, 10). From his analysis of the "threshold" (which even today is perceived as an invisible reality) and of the "door" that is so "hinged" that it lies between "outside" and "inside," he comes to the conclusion that these last two meanings are complementary and cannot be reduced to entities or

Ivan Illich

Plato's Motherly Space

It is very difficult to evoke the sense of "space-as-substance" among modern city-dwellers.[10] They cannot perceive space as "stuff"; they cannot imagine smelling or feeling it. It is therefore consoling that for Plato "to express [himself] in clear language on this matter [space] will be for many reasons an arduous task." The statement occurs in *Timaeus* (49-52), which also deals with the foundation of Athens through the agency of Pallas Athene.

Timaeus describes the second of the three great principles that make understanding possible: "First, there is that which is in the process of generation; second, that in which generation takes place; and third, that of which the thing generated is a semblance." The first, in our case, will be the tangible

---

categories which would embrace both at the same time. They did not, therefore, originally "mean" two kinds of "space." See also **Meister** and Haight. **Peuckert** studies a number of folk stories from various parts of Europe in which travelers are pursued by evil spirits and escape by reaching an "inside" in time. These stories are very common and allow him to distinguish two kinds of frontiers between inside and outside: the boundary of the fields of the villages and the eaves-drip (or drip-edge, German: Traufe) of their home. Village and house are both shelters, albeit of a different order. **Schmidt** makes the same observation examining rituals and their reflection in poetry. Up until the nineteenth century, the term "no man's land" referred to that other "whereabouts" beyond hallowed space. Only in World War I did it come to mean the "space" between the trenches. See also **Norden** and **Trumbull**.

10. The use of Sanskrit enables Indian philosophy to clarify the differences between space as matrix, on the one hand, and space as medium of localization, on the other. According to **Filliozat (1969)** seven *dravya* are distinguished quite commonly: substances or "kinds of stuff" that can acquire *gunas*, that is, properties. Four are the elements: water, fire, earth, and wind; the fifth is *rata* (time). The other two are mixed up in our concept of "space" but are distinguished in Sanskrit by two words: one is *akaça*, the void which contains all, "neither long nor short, without form, taste, smell, dimension, a mere container" and the other is the "dyssymmetric void," the organized void within which objects are placed.

reality of Athens, the second the founded space within which it comes into being, and the third the *templum*, the idea of the town which "is prior to these others, and known only to God and he of men who is the friend of God."

At this point in the dialogue, Timaeus is concerned only with the second of these principles, "the nurse of all things that are generated," the "receptacle . . . that we may liken to a mother." "Mother" in Greek, as in older forms of English, is synonymous with "womb," not with "woman." "She is the natural recipient of all impressions, and is stirred and informed by them, taking different appearances at a given time." "Wherefore, that which is to receive all forms should have no form itself, as in the making of perfumes, where they first contrive that the liquid substance which is to receive the scent shall be as inodorous as possible. . . ." "Therefore the mother and receptacle of all created and visible and sensible things is not to be termed either earth or air or fire or water . . . but an invisible and still undetermined thing . . . which, while in a mysterious way it partakes of the intelligible, is yet most incomprehensible." Of this "receptacle and nurse of all generation," Timaeus continues, "we have only this dreamlike sense, being unable to cast off sleep and determine the truth about it" because it exists "only as an ever-fleeting shadow," even though out of it all tangible things are generated. In these delightful lines Plato still speaks of the image-pregnant stuff of dreams and imagination; he speaks as an early philosopher, however, as one who still has the personal experience of living and dwelling in precategorical "founded" space.

With Aristotle, space ceases to be understood as such a "stuff." Plato's "receptacle" (*hypdechomene*) is transformed by Aristotle into one of the logical four "causes" of existence and identified with matter (*hyle*). Aristotle lays the foundation for a perception of space on which Western space perception ultimately builds, space not as receptacle but as expanse. Beginning with Aristotle, the "ideal city" becomes a juridical fiction.

Up until Plato's time and on occasions, as we shall see, even later, the *invisible* city was a tough reality. Once inaugurated, a city is almost impossible to get rid of. The ideal space of a town cannot be eradicated; it survives the leveling of its walls, the burying of its buildings, and the enslavement of its inhabitants. After Scipio, the Roman general in the Third Punic War of 146 B.C., had desolated Carthage, he had by no means completed its destruction. He had not undone its foundations until Carthage was "plowed under." Its sacred furrow had to be reversed: those clods that in the foundation ritual had been carefully heaped toward the inside had to be returned to the outside. When he ordered the un-plowing, Scipio probably thought of Achilles, who dragged Hector's body thrice around Troy to "clean" (*lustrare!*) the spot, thus making Troy disappear. Only when a city's soul has been snuffed out is its claim to tribute extinguished and the wilderness allowed to swallow up the site.

The Bulldozed Space

Scipio's plowing under of Carthage evokes an eerie feeling for anyone who has ever lived in a Rio de Janeiro *favela* and seen bulldozers sweep down on its shanties. Two sentiments are incongruously entwined at this moment: a sense of déjà vu and a shock at the unprecedented nature of the confrontation being witnessed. Carthage and Rome faced each other as homogeneous enemies, two entities grown out of the same kind of stuff. But when the bulldozers come escorted by the police, two unlike entities meet—shanties grown out of dwelling space versus aggressors from a world constructed on the drawing board. One would underestimate the violent heterogeneity, the radical unlikeness of these two entities if one were to compare it to a science-fiction encounter between three-dimensional humans and visitors from multidimensional space.

Yet it is due perhaps to this incongruity that a *favela* once established does not go away just because its site has been bulldozed. Within weeks, even overnight, the same *favela* will be there again. If you watch in the evening after sunset, a hundred families will climb over the barricades carrying poles, mats, and infants. By dawn, dozens of women will emerge from the wobbly shelters to line up—as they have always done—and fill their buckets at the nearby spigot. For the most part, they will be different people from those who have just been carted away. Not the same people, perhaps, but the same *favela* has returned to its visible life. Certain towns, like Jericho, have had several lives. The bulldozer is as powerless to eliminate invisible space today as were the Roman legionnaires of antiquity. But cement can crust it over. When a parking lot is built or public housing rises on the site, the squatter can return no more. The ancients believed in their power to undo ritual space; they knew that it was a social creation. Architects can only condemn it and bury it under cement. And, as the world is cemented over, dwelling space is extinguished. It survives only in cracks and niches. Most people are forced to acquire costly space in which they cannot dwell.[11]

---

11. It would be a grave mistake to generalize from Etruscan foundation rituals as though they were the model according to which dwelling space is ritually created by all cultures. The rituals described here should be seen as only one ideal type through which social space can be brought into existence and maintained. In certain African traditions, beautifully described by **Zahan**, I have the impression that social space is cultivated as the result of the personal experience of initiation. The initiatory way into the sacred woods and the ritual discovery of one's one "inner existence" are expressed in the communitary building of house and village. This example might be seen as the inverse of the Roman procedure,

Ivan Illich

In-discrete Space and the Nightmare

The bulldozer incorporates the *favela* into the modern metropolis. It breaks down the distinction between outside and inside space.[12] It incorporates discrete vernacular space

---

through which the templum, made visible in the city, comes to be experienced as an inner reality. **Lebeuf** reports from the Congo a "creation of space" that is the result of heaven and earth growing together, as the right and left parts of the house are carefully built so as to rise, inch by inch in harmony with each other. **Roumeguere** describes the distinct stages of an initiation ritual, in each of which a new revelation of the body's significance associates the young man or woman with a different sphere of outside social realities. **Niangoran** stresses even more than Zahan that some African dwelling-spaces are the result of each generation's initiation and therefore are time-bound. They are constantly in the process of decaying and must be reconstituted. Nicolas reports that the sacrificial victim is "split" to "make" new space. The space-creating spirit is ever at work as a zigzag line, representing the motion of water, word, and dance. See **Griaule**, 12, 18ff., 138ff., on the "Nummo pair of twins, who are water." Space seems never to be "sealed off."

12. The *OED* gives the following definition for *indiscrete*: a. 1608; unseparated. 1. "not distinctly distinguishable from contiguous objects or parts." The *door* or *gate* loses its meaning when it ceases to be the point of encounter between two worlds.

> Türen gehören der Vergangenheit an ... und wie soll es Türen geben, wenn es kein Haus mehr gibt... die Tür war ein Eingang zu einer Gesellschaft von Bevorzugten, die sich dem Ankömmling, je nachdem wer er war, öffneten oder verschlossen, was gewöhnlich schon sein Schicksal entschied. Ebenso aber eigneten sie sich auch für den kleinen Mann, der aussen nicht viel zu bestellen hatte, jedoch hinter seiner Tür sofort den Grossvaterbart umhängte. Sie war darum allgemein beliebt und erfüllte eine lebendige Aufgabe im allgemeinen Denken. Die vornehmen Leute öffneten oder verschlossen ihre Türen, und der Bürger konnte mit ihnen ausserdem ins Haus fallen. Er konnte sie auch offen einrennen. Er konnte zwischen Tür und Angel seine Geschäfte erledigen. Konnte vor seiner oder einer fremden Tür kehren. Er konnte jemand die Tür vor der Nase zuschlagen, konnte ihm die Türe weisen, ja er konnte ihn sogar bei der Tür hinauswerfen ... diese grossen Zeiten der Türen sind vorbei ... wer hat je wirklich "einen hinausfliegen" gesehen ... und vor seiner eigenen Tür zu kehren ist eine unverständliche Zumutung geworden ... nur noch freundliche Einbildungen die uns mit Wehmut beschleichen, wenn wir alte Tore betrachten. ... Dunkelnde Geschichte für ein Loch, das die Gegenwart vorläufig noch für den Zimmermann offen gelassen hat.

H_2O and the Waters of Forgetfulness

sui generis into non-discrete, in-discreet, homogeneous, commercial space. Each dwelling space is the stuff for its own unique kind of housing. Non-discrete space must first be created and then be allotted to garage people in units of flats. The bulldozer can be taken as the symbol of societies like ours, of societies that exist in *indiscrete* space. Such exceptional societies cannot be compared with any that have previously existed: preindustrial societies could not have existed in homogeneous space. The distinction between the outside and the inside of the body, of the city, of the circle was for them constitutive of all experience. The dissymmetric complementarity of the exterior and the interior, of right and left, male and female was a root experience. The homogeneous space which transcends this distinction is, historically, a new kind of experience. It constitutes a continuum which was formally not experienced, a continuum that is neither interior nor exterior, neither right nor left. In societies that can experience this geometrical continuum,[13] the "exterior" and the "interior" are just two locations within one kind of

---

(R. Musil, "Turen und Tore," in *Unfreundliche Betrachtungen aus: Nachlass aus Lebzeiten.* Rororo *Werke* 7:504-06.)

13. From the time Rudolf zur Lippe first began working on his doctorate with **Theodor Adorno, his** primary subject has been the "geometrization of the human being." He coined this term mainly to refer to an attempt (around 1660) to deduce the laws governing human life from the observations of nature made in a Cartesian perspective. Lippe focuses on the courtly celebration and the personal interiorization of this new perception. For Lippe this geometrization finds its expression in new ways of gardening and of swordplay, in new forms of etiquette and military discipline as well as in building. His major work (**Lippe 1977**) deals with the use of dance and swordplay as rituals consciously used for this purpose. In several expositions and their corresponding catalogues, he highlights the use of architecture and military drill for the same purposes. Insofar as I am searching for the origins of the monopoly of indiscrete (geometric) space on social imagination, my argument is based on my understanding of Lippe.

"space." "Home" and "abroad," "dwelling" and "wilderness" are nothing but regions or areas or territories selected from the same expanse. In this bulldozed space people can be located and given an address, but they cannot dwell. Their desire to dwell is a nightmare.[14]

Italo Calvino has described this nightmare in his *Invisible Cities*. He tells of Marco Polo's visit to the court of Kublai Khan. Polo tells his host about the stuff of the towns through which his imagined travels have led him. Calvino has Marco Polo describe the sickening helplessness that he experiences as a man accustomed to traveling in three-dimensional space, when led through dreams of cities, each generated by a different "stuff." Polo reports to the Khan on dreams of space with a pervasive taste of "longing," on space made up of eyes, of granular space that jells into "names," of space that is made up of "the dead," space that constantly smells of "exchanges" or "innovation." Marco Polo reports on these nightmares for the benefit of his host and ends with the following entry: "Hell—if there be such a thing—is not tomorrow. Hell is right here, and today we live in it; together we make it up. There are only two ways to avoid suffering in this Hell. The first way out is easy for most people: Let Hell be, live it up, and stop noticing it. The second way is risky. It demands constant attentive curiosity to find out who and what in the midst of this Hell is not part of it, so as to make it last by giving space to it." Only those who recognize the nightmare of nondiscrete space can regain the certainty of their own intimacy and thereby dwell in the presence of one another.

Inner and "Outer" Space

14. "Space is nothing but a 'horrible outside-inside.' And the nightmare is simple, because it is radical. It would be intellectualizing the experience if we were to say that the nightmare is the result of a sudden doubt as to the certainty of inside and the distinctness of outside." (Gaston **Bachelard**, *The Poetics of Space*, 218).

"Taking up space" and "giving space" are interwoven into the art of dwelling. To dwell means to draw out of the city's matrix a dreamlike stuff, to spin threads from it, to use these to form a warp harnessed to the city's *templum*, and to weave action into this warp.

Even our thoughts must be woven into this warp, unless we want to disconnect them from the fabric of life. I cannot think in harmony with my imagination without implying such a warp of imaginary space. As soon as I say that I have "come to know something," I have already kept my distance from it. I have "looked" at it, "searched" for it, "figured out" the "right angle" to "approach" it, "reached out" for it, and finally "grasped" it. All these verbs that allow me to describe the "process" and "progress" of my thinking are, of course, spatial metaphors, and they all refer to space that is within me. When I use any one of these expressions, I am aware that the space which I experience between myself and the world that I have come to know is not "in" the same kind of space, "in" which I perceive the things around me. I am told that "in my mind" is a systematically misleading expression and that I should dispense with it as thoroughly as I can.

I cannot follow such well-meaning advice. Insight leads me to the perception of an in-side. Starting from the ritually drawn perimeter (the furrow, the skin, the social domain) each age creates its own dissymmetric complementarity between these two *sides*. The inside and the outside are spun—one clockwise, the other counterclockwise—from the matrix of each culture. By insisting on interior "space" I defend myself against the geometrization of my intimacy, against its reduction to an algebraic notional equivalent to an exterior space that has been reduced to Cartesian dimensions.[15] Such an intrusion would allow "nondiscrete" space

---

15. For a classical analysis of mental space-time and its perception, see **Alexander**, 1:133-35. For Samuel Alexander (1859-1938), metaphysics deals with comprehensive features of experience that lie outside the pur-

to flay my intimacy and thereby render it extinct, as Carthage was plowed under by Scipio.

I equally refuse to give geometry a monopoly over things that are not part of my intimacy. Most cultures have eyes that see "out there" realities that cannot fit into the formal continua of mathematics and physics. Neither the Greek gods nor the ghosts of popular culture nor the elementary spirits of fire and water and air, that according to Paracelsus in his treatise on nymphs, sylphs, pygmies, and salamanders (Blazer) inhabit the elements, can dwell in such Cartesian continua. Geometry is not a spindle that can draw out yarn for the shuttle with which my imagination weaves.

Elusive Waters

The water that we have set out to examine is just as difficult to grasp as is space. It is, of course, not the H_2O produced by burning gases nor the liquid that is metered and distributed by the authorities. The water we seek is the fluid that drenches the inner and outer spaces of the imagination. More tangible than space, it is even more elusive for two reasons: first, because this water has a nearly unlimited abil-

---

view of the special sciences. They can be understood, by inspection, as pervasive features of the world that we study empirically. Space-time is the universal metaphysical matrix, the titan out of which different levels of organization emerge. These "emergents" cannot be explained but must be accepted with "natural pietry" (a phrase borrowed from Wordsworth). The space-time matrix is thus pregnant with the matter, life, and mind that appear from the observer's perspective. "Body" is the external view of nature as unified in the particular historical perspective. "Mind" is the "idea" of the distinctive internal quality that is implicit in such a perspective. Alexander's thought is the independent equivalent in English of the best that German phenomenology has to offer. He was proudly conscious that he could be accused of having "erred with Spinoza." See **Jammer** (first chapter) for the transition from Platonic to Aristotelian space. The concept of a "systematically misleading expression" that produces an error in category is taken from **Ryle**.

ity to carry metaphors and second, because water, even more subtly than space, always possesses two sides.

As a vehicle for metaphors, water is a shifting mirror. What it says reflects the fashions of the age; what it seems to reveal and betray hides the stuff that lies beneath. On the Wilhelmshoehe near Kassel, a German baroque prince has surrounded his castle with an English garden that solicits his waters to spill all that they know. As a man of his age, he even developed a taxonomy of water's secrets. His architects decided where in this park waters were to be clear or sparkling or deep or open or dull. In the woods they gush and mumble and ebb and swell, and in the meadows they meander and dally and trickle down in the grotto from the roof. There are niches and walls that are misty or dewy or wet. His waters tease and seduce; they threaten to drench and even to drown you. The prince's waters are there to amuse a whole court.

However, it is not this everchanging surface of water that makes it so difficult to explore the historical "stuff." It is the deep ambiguity of that stuff itself that makes it as elusive for us as space was incomprehensible for Plato. Water remains a chaos until a creative story interprets its seeming equivocation as being the quivering ambiguity of life. Most myths of creation have as one of their main tasks the conjuring of water. This conjuring always seems to be a division. Just as the founder, by plowing the *sulcus primigenitus*, creates inhabitable space, so the creator, by dividing the waters, makes space for creation.

The Division of Waters

In Maori myth, creation starts in the womb, in which the waters fuse. The firstborn wedges himself between mother and father, whom he thereby separates from each other; from the blood the separation draws out of the womb the world is made. In the Rig-Veda, Indra, the god, is the germ of the waters. He rises from the dark lap of the limitless flood, like a fiery column, as the waters that encircle him glow and sing.¹⁶

In the first chapter of Genesis, on the second day, "He said, let a vault arise amid the waters, to keep these waters from those; a vault by which He would separate the waters beneath it from the waters above. And so it was done.¹⁷ This vault He called sky. So evening came and morning came and a second day passed." The waters rebelled against this separation. The chaos refused to make space for creation. The waters destined to be up high refused to leave the embrace of the waters resting below, and they embraced each other more closely. According to the midrashim (known to Philo, Origen, Jerome, and Albert the Great) the second day was the day of God's cosmic struggle. He almost gave up the work He had begun. Only the foreknowledge that a remnant of Israel would remain faithful made Him return to his job. This one day He did not say "And it was good" because He knew the waters were weeping on account of their separation, and seeing their tears He too was sad. Some say that He spread his own mantle between the waters; others say that for this

16. **Kramrich** comments on the pertinent passages from the Rig-Veda: "As bull he generates them, and as child he sucks them and they lick him. . . . He enters them who do not eat and are not deceived, the restless daughters of heaven who do not dress and are not naked . . . having one lay in common . . . the seven melodies receiving one germ in common . . . and he is the womb of the mother."

17. **Ginzberg** 1:13–18 and his notes and bibliography (see his footnotes to *Legends on the Pentateuch* 5, notes 48, 49, 54). It is noteworthy that Rabbi Akiba warned his pupils not "to shout—water!—whenever seeing a vision of crystal around the throne of God" (Hag 14b). On the "division of humors," the separation of tears into those of laughter and those of pain, start with **Hvidberg**. Jewish teaching and Christian theology have interpreted Genesis 1 as revelation about Creation from "Nothing" and made "stuff" as much as form contingent on the word of God. This theological idea has had a deep influence on western philosophical thought: see **Sertillanges**.

purpose He used a shard. All sources agree that He sealed this "firmament" with his own ineffable name and appointed a special angel to watch over the integrity of the sky. This angel, appointed eons before the other one who stands at the door of Paradise, holds the great seal, and each time a Jew's curse rends the mantle, the angel is there to repair it. Only when He had finally succeeded in splitting the waters could He set out to create the earth.

To keep one's bearing when exploring water, one must not loose sight of its dual nature. In many African languages the word for the "waters of the beginning" is the same as that used to designate twins. Dream-water there is two-faced. The flood, the blood, the rain, milk, semen, and dew, each of the waters has an identical twin. Water is deep and shallow, life-giving and murderous. Twinned, water arises from chaos, and waters cannot be but dual.

Water's Dual Nature: Purity and Cleanliness

One very special way in which the dual nature of water shows is water's ability to purify as well as to clean. Water communicates its purity by touching or waking the substance of a thing and it cleans by washing dirt from its surface.

The substantive purity that water radiates is not my theme; it is, rather, the other side of the subject I am pursuing. I wish to focus on the ability of water to wash and must be careful not to be misled and distracted by its purity. My theme is the power of water to clean, to detach what sticks to people, to their clothes or their streets. The power water has to penetrate body and soul and communicate to them its own freshness, clarity, and purity is another theme with an altogether different history.

The distinction between purification and cleansing is obvious yet difficult to clarify. The late archaic transformation of *miasma* in Greece, followed by the gnostic tradition and baptismal theology, has jumbled purifying blessing and detergent scrubbing under the emblem of "water" that determines modern sensibilities.[18] In our century psychology and the religious sciences have continued this jumbled tradition and, consequently, the discussion of the power of water to detach and purge filth has been left to hygiene and engineering. As a result, the symbolic functions of ablution and laundry, insofar as these are distinct from purification, have been little explored.

Purification is by no means a process for which water is always needed. Water is often used in this process even though purification is also performed by other means: blood is used, but also incantations, noisy processions, ecstatic dances, the imposition of hands, induced trances or dreams, the wearing of amulets, fumigations, or contact with fire. However the purity that water restores or confers has a special connotation of freshness and transparency that transforms the innermost being and so it is often associated with re-birth.

A reflection on vocabulary is helpful to clarify the difference between purification and cleaning. Purity refers to a quality of being. Even when this quality appears on a being's surface, it is perceived as the manifestation of something deep inside. Its beauty can be lost only through a corruption at the being's core. There is no one word to say what is then lost. The loss can be expressed only with a negative compound: we cannot help but say "impure."

In contrast with this negative reference to the condition that calls for purification, Indo-Germanic languages possess a rich register for referring to the conditions that require

18. "Miasma", see Pokorny (*Indogermanisches Woerterbuch*): +mai- (=moi-?) beflecken, beschmutzen; Anglo-saxon *mal*, n. "Fleck, Makel"—Old High German. The same meaning for *meil-* and possibly its equivalent in Lithuanian (*Sumpfwiese* or marsh-meadow), Kluge in Mhd. mal, sbstv Onians, 585; discolored spot in Old English; spot or blemish on the human skin after the fourteenth century.

H_2O and the Waters of Forgetfulness

cleansing. "Miasma" that can be washed away is given in bold and direct terms as something that sticks to the skin, such as soil (soiled), shit (a word that comes from the same root as "dirt" and "dirty"), foul things (filthy), dung (from Old Germanic *quat*, *Kot*), glue or sap (*sucio*, *sudicio*) or mud (Irish *loth*, from the same root as Latin *lutum*). These are all earthy things that water washes away. It acts as a solvent (it ab-solves), detaches these leftovers of past activities and disengages the person from an encumbrance. Nor is it only the condition requiring such cleansing that is directly expressed in our languages; what water itself does is described by means of several different verbs: fingers, face, and mouth are rinsed; clothes are laundered; the body and also the feet are bathed when they are washed.

In one and the same ceremony water can sometimes both purify and clean. This action is most evident in the washing of the dead. The custom is attested as far back as Homer, and it has remained, well into the twentieth century, a common feature of Christian, Jewish, and Muslim funeral ritual from Morocco to the Urals. The task has been elevated by the church to the dignity of an act of mercy. Ignatius Loyola imposed it on his novices before he would accept their vows as Jesuits. But apart from such masculine heroism, it has remained an act performed primarily by old women, widows, and semi-witches—not infrequently by the same women who also wash the newborn. Washing the newborn or the dead is fraught with dangers that women face better than men. Before starting her cleansing, the Jewish woman places a kerchief on the face of the corpse; the Russian woman bows deeply and asks the dead to forgive her for stripping his remains. The ceremony is performed mainly to divest the corpse of an aura that attaches to it, one that should not go along with the dead into the grave.

Great care is taken to dispose of the water used on such occasions in such a way that the corpse will not pick up this aura again. Only bodies so washed will not stay glued to their environment, will not remain prisoners of this world and haunt those who are still alive. What for the dead man or woman is "ablution," "absolution," delivery from burdensome soil and dirt is, for the living, a purification of their dwelling space corrupted by death.

The Waters of Lethe Wash

Once the dead man has been washed, he can set out on a journey. All Indo-Germanic pilgrims—Greek, Indic, Nordic, and Celtic—cross the same funeral landscape on their way to the beyond, and the mythical hydrology on that route is the same: at the end of their journey they reach a body of water. This water separates two worlds: it divides the present from the past into which the dead move. This other world does not have one common fixed location on the mental map of Indo-Germanic myths; it may be located below the earth, on a mountain top, on an island, in the sky, or in a cave. However, this other world is always a realm lying beyond a body of water—beyond ocean, river, or bay. In some regions one crosses this water on a ferry; in others one must wade or swim. The slow, flowing waters the traveler crosses are everywhere emblematic of the stream of forgetfulness; the water has the power to strip those who cross it of memories that attach them to life. The sleepy beating of the head in the *threnos* with which the mourning women lull the heroes of Thebes into their last sleep reminds Aeschylus of the monotonous beat of the oars across the river Acheron.[19]

This river, which sums up recollections, detaches memories, detaches from the dead those deeds that survive them,

---

19. Ninck, 122, note 2: "This picture offers us a profound glimpse into the character of sorrow. . . . The lullaby for the other side must be heavier and duller, as though it were putting one to sleep." Marringer surveys what we know about the meaning given to water in prehistoric times; water was both "fertile" and the "shore" for the other world.

came to be called "Lethe" by the Greeks. Just as the Egyptians, for whom the Nile had been the divide between the two kingdoms, placed the reign of shadow on the western bank toward the horizon where Heaven and Earth are fused, so in late antiquity this body of water was located in far-off Galicia. During the Middle Ages the poor souls on the way to purgatory had to cross the Atlantic Ocean to reach the fabulous island of Saint Patrick, shown to the northwest of Cabo Verde until late into the fifteenth century.

Bruce Lincoln has shown that there is yet another common feature in all Indo-Germanic mytho-hydrography. What the rivers or beaches wash from those who cross them is not destroyed. All mythic waters feed a source that is located on the other side. The streams carry the memories that Lethe has washed from the feet of the dead to this well thereby turning dead men into mere shadows. This well of remembrance the Greeks called "Mnemosyne." In her clear waters, the residues of lived-out lives float like the specks of fine sand at the bottom of a bubbling spring. Thus a mortal who has been blessed by the gods can approach this well and listen to the Muses sing in their several voices what is, what was, and what will be. Under the protection of Mnemosyne, he may recollect the residues that have sunk into her bosom by drinking from her waters. When he returns from his journey, from his dream or vision, he can tell what he has drawn from this source. Philo says that by taking the place of a shadow the poet recollects the deeds which a dead man has forgotten. In this way the world of the living is constantly nourished by the flow from Mnemosyne's lap through which dream water ferries to the living those deeds that the shadows no longer need.

Mnemosyne's Pool of Reflection

Mnemosyne is one of the Titans. She appeared when the sky still rested in the arms of the earth, when Uranus shared the bed with broadhipped Gaia, an eon before the Olympian gods were born. The *Hermes* calls her the Mother of Muses. Hesiod remembers her flowing hair as she stretches out to engender with Zeus her nine daughters. It is she who adopts the son of Maya, the "shamefaced" or "awful" nymph, and thus makes him the son of two mothers. She provides Hermes with his two unique gifts: a lyre and "soul." When the god Hermes plays to the song of the Muses, its sound leads both poets and gods to Mnemosyne's wellspring of remembrance. Hermes is both messenger and guide of the Gods. Even the immortals must draw on the waters of his titanic mother if they want to remember. The appearance of Mnemosyne among the Titans is crucial for our history of water; it tells of water before there were gods. Being placed among the Titans, a cosmic element—water that washes—became the source of remembrance, the wellspring of culture, and acquired the features of woman.

However, the first woman of oral tradition is forgotten when the oral transmission of epics ceases. The classical poet of Greece no longer has need of recollections from a "beyond." His sources are frozen into texts. He follows the lines of a written text; the epic river that feeds its own source is remembered no more. Not one Greek city has preserved an altar dedicated to Mnemosyne.[20] Her name becomes a technical term for "memory" now imagined as a page; the stuff of

20. **Vernant 1959** emphasizes that those who approach the muses by returning to the source whence they have sung from the beginning do not find a "precedent" for the present; they leave the temporal frame and listen to the bubbling which has gone on without interruption since the beginning: ab initio, they come to understand the "happening." See also **Eliade 1963**, ch. 7, on the mythology of memory and forgetfulness. Eliade underscores that Lethe's function is turned topsy-turvy when the Indian-gnostic concept of metempsychosis (transmigration of souls), mediated through Plato, makes its waters of forgetfulness into the agent by which the soul returning to earth is washed of its memories of previous lives. When Lethe acquires this new function of washing the soul des-

H_2O and the Waters of Forgetfulness

memory turns from water into a shard. Written language, which has fixed words on clay tables, acquires more authority than the re-evocation of fluid, living speech.

Many types of "writing" had been known before, but all of these scripts were like railings, landmarks, or arrows guiding the flux of speech in the right direction. Pictograms or ideograms did not have the exclusive technical function of fixing sounds, as pronounced, so that they might be voiced again precisely in the same way but at a later time and by someone else.

Before epic tradition was recorded, before custom could be fixed in written law, thought and memory were entwined in every statement; the speaker had no way to imagine the distinction between thought and speech. Voice could not be stocked, it left no dregs or grounds. Solemn composition had to fit the rhythm of the hexameter, stressed by the beat of the lyre strings. Consciousness, lacking the metaphor of the alphabet had to be imagined as a stream full of treasures. Each utterance was like a piece of driftwood the speaker fished from a river, something cast off in the beyond that had just then washed up onto the beaches of his mind.[21]

---

tined for reincarnation, anamnesis (remembrance) becomes the rediscovery of past selfhood and ceases to be the plunge into collective mythos.

21. **Notopoulos** finds Mnemosyne, as she occurs when Greek literature was first committed to writing, to be the personification of a vital force later forgotten. In this way Memory was conceived as a means in the process of creation; Mnemosyne was identified with the inner source of inspiration bubbling at the speed of the lyre. What bubbles up from with Mnemosyne is "thought" and "verse" and "remembrance" in a connection so close as to be understood, of necessity, as the muses' song. Remembrance, in oral epics, is the echo of an inner voice. The poet's words were to inspire memories, to guide to Mnemosyne's well. The poet did not strive to have his words stored, remembered. In *De Bello Gallico* Caesar reports on the Druids "Who learn by heart a great number of verses . . . and they do not think it proper to commit these verses to

The idea of dessicating and freezing the voice of the muses must have been deeply offensive to the contemporary mind. In *Prometheus Bound*, Aeschylus reflects the strong prejudice against the alphabet words common in the fifth century. Among the gifts Prometheus brought to mankind, as a culture hero, "was the combining of letters, creative mother of the muses' art, wherewith to hold all things in memory." For presuming to lock up the Muses in script, Prometheus was cruelly punished by Zeus. He had engendered his daughters in the pond of Mnemosyne so that they might bubble and flow, not dictate. Plato in the early fourth century was anguished by the effect the alphabet was having on his pupils. Their reliance on silent, passive texts could not but narrow the stream of their remembrance, make it shallow and dull.

The Aqueduct and the Alphabet Dry Up Mnemosyne

During the very same centuries during which the letters transformed Greek attitudes to memory the Greeks also engaged in what Dodds has called "the moral education of

---

writing. I believe they have adopted the practice for two reasons—that they do not wish their wisdom to become common property, nor those who learn it to rely on writing and so neglect the cultivation of memory." In his comment on Plato's *Phaedrus* 275 a–e, Notopoulos emphasizes that for Plato recourse to Mnemosyne was a reaching after originals free from symbolism, while remembering in the context of writing is always a reference to an eidolon, a lifeless image. For details see **Illich**, *Phaidros und die Folgen*, ch. 2–3.

Notopoulos insists on the extreme ambiguity of Plato in the face of the written word: "Plato as the author of the dialogues is perhaps the most gifted child of written literature . . . but wishing to preserve the memory of his master he launched on the composition of the dialogues, like the oral poet in aim but unlike him in the means of expression." For the rebirth of the topos of mnemonic waters in early patristic times, see **Lewy**. For an introduction to orality and literacy, see **Ong**. For the state of the art in distinguishing mnemosyne from mnemonic storage, see **Peabody**.

their Zeus."²² If we compare Homer's version of the Oedipus saga with the one familiar to us from Sophocles, it is clear that in the first version Zeus did not care about "justice." Oedipus continues to reign in Thebes even after his guilt is discovered and is eventually buried with royal honors after being killed in battle. There is no trace in Homer of the belief that pollution was either infectious or hereditary. In the fifth century version that Freud has made familiar, even to those who have no idea where Greece might be, Oedipus becomes an outcast crushed under the burden of his guilt. Guilt has penetrated his being, where neither rain nor sunlight can wash it away; it is experienced not with shame but with anxiety. It is more terrifying now because the guilty suffers a new uncertainty: he seeks and cannot reach its source. His guilt could be the result of a chance contact or else inherited through the forgotten offence of some ancestor. The pollution of early, pre-alphabetic Greece belonged to the world of external events. It operated with ruthless indifference to motives or justice. The removal of polluting miasma, was performed in Homer's world by simply washing off what had gotten stuck to the skin. Before the fifth century Dodds does not encounter explicit statements that clean hands are not enough, that the innards had to be clean. In the late archaic period anxiety about individual purity quickly became an obsession in Greek religion. The tales about Orestes and Oedipus were recast as horror stories of blood guilt, and purification became the main business of Delphi. As the

---

22. Purity is just as historical as cleanliness. Both—to invent a metaphor—stick to the skin. Cleanliness says something about the outside of the skin and purity says something about what is beneath it. I suspect that the senses of purity and of cleanliness can change historically with considerable independence one from another. This is what **Parker** in his recently published rich study suggests. My thinking is strongly influenced by **Dodds 1951**, especially ch. 2, "From shame culture to guilt culture," also **Zucker, Meyer, Latte, Snell**.

waters of memory were silenced, increasingly ritual waters were used to communicate their own purity to the guilty man. The action of washing something off the person became the metaphor, while "aspersion"—the physical contact with water's intrinsic purity—was the operation that now brought reality to the act.

Rome's In-discrete Water Works

During the same period of time that it took to replace a fluid memory with a fixed storehouse for past utterances and to emphasize the morally purifying agency of water, the status of water within city space also changed. Cities became dependant on water brought in over aqueducts that pierced the city wall. Early settlements had depended on rivers, ponds, and enclosed springs from which water was occasionally channeled to a nearby temple or palace. The art of well-digging was being perfected after 2500 B.C. Egyptians learned to "mine" water, to increase the output of their wells by driving horizontal tunnels into the strata at the bottom of the hole they had dug. In Palestine deep passages were burrowed through to subterranean springs outside the city and led to ponds accessible by deep stairways. Mycenae was the first European city to get some of its water delivered by tunnel. But only during the course of the seventh and sixth centuries B.C. did the aqueduct become an accepted feature of the landscape. Water was brought from great distances to Nineveh (fifty-five miles), Troy (seven miles), Athens, Corinth, Megara, and Samos (a one-kilometer tunnel!). Until 312 B.C. (441 *ab urbe condita*), Romans were contented with the Tiber, some springs, and a few enclosed wells. By A.D. 97, Rome had become a city of one million inhabitants. Nine major aqueducts with a total length of approximately 250 miles brought 100 gallons of water per capita into the city. How much water this is, one can grasp only by comparison.

H_2O and the Waters of Forgetfulness

London, Frankfurt, and Paris had eight-tenths of a gallon per capita in 1823 and approximately ten gallons per capita in 1936.[23] Rome in A.D. 100 used ten times this amount of piped water. The man who was in charge of the water system that year, Sextus Julius Frontinus, has left us a detailed description of the way the Roman aqueducts worked. One-fifth of the water went straight to the emperor, another two-fifths to the city for its 591 fountains and dozen public baths. Frontinus is proud of this public display of the government's might: "With such an array of indispensable structures carrying so many waters, compare, if you will, the idle pyramids or the useless, though famous works of the Greeks."

Rome's glory was the ostentatious domestication of Mnemosyne both through the codification of public memories in Roman law and through the piping of city water. Rome's architects picked up a source in the mountains, chan-

---

23. Aqueducts brought about 300 liters per day per capita into imperial Rome; another 150 liters were lost or stolen on the way to the fountains. Beginning with the second century B. C. numerous aqueducts also started to feed Tarragon, Segovia, Nimes, Ephesus, Antioch, and Carthage. To have this amount of water per capita in the household and garden would seem quite satisfactory to most modern Americans. Today industry consumes most of the water, from five to ten times as much per capita as households. But it would be a grave mistake to take this water consumption as an indication of Roman "hygiene." **Mumford** (ch. 8.2) concentrates on this issue:

> Just as our expressways are not articulated with the local street system, so the great sewers of Rome were not connected to the eroded tenements at all.... Where the need was greatest, the mechanical facilities were least.... The same uneconomic combination of refined technical devices and primitive social planning applied to the water supply.... Water and slops had to be transported by hand in the high tenements, just as they were transported to equally high tenements of seventeenth-century Edinburgh.... For all its engineering skills and wealth, Rome failed miserably in the rudiments of municipal hygiene.

Mumford then quotes from Rudolfo Amadeo Lanciani's account of his excavations (1892) of the ancient Roman *puticuli*, the open pits into which refuse and dead slaves by the tens of thousands were thrown and which still form huge gelatinous masses beneath Rome.

neled the water unmingled into the city, and chose for each one of the waters the stories it should tell in the city. Each fountain was hewn into marble and displayed as a work of art. The artist used the water to give sparkle to the tritons and nymphs of his invention, and the Senate chose the street crossing to exhibit its power over that water flow. By making a mountain source into a city fountain, Rome broke the magic circle that the founders had plowed around city space. The water that splashed in the fountains of Rome was non-discrete water, at home neither inside nor outside.

But Rome was not just any city; she was THE city, the *urbs*. As the water was leveling its furrows, Roman space began to explode from within. The *urbs* became the center of the *orbis*; this unique city space sprawled beyond its limits to engulf the *Orbis Romanus*. Space became "catholic," that is, universal. The city of Rome offered citizenship within its uncircumscribed space. With indiscrete space and "catholic" water the "human" being acquired new meaning. Rome left its well-watered space as a heritage that survived its aqueducts, which were all destroyed by the nordic invaders. For 800 years, from the sixth century to the fourteenth century, Rome lived again, as in the early republic, from the waters of its wells and river, at the time when the Church, through its baptism of water, incorporated Europe into the City of God. While only the elect in the provinces had been allowed to become citizens of Rome, under the Church's regime all people were called to abandon their "paganism" and to be "washed" into catholic space.

Harvey Invents Circulation

Other societies have engaged in the display of piped water, but two factors distinguish them from the waters of Rome. First, non-Roman displays of water did not have the same political purpose; and second, the piped waters that flowed

H_2O and the Waters of Forgetfulness

into fountains and baths were not wasted, but carefully put to use. Muslim princes from Granada to Isphahan and Agra who had relished fountains were careful lest their flowers and gardens should lose a single drop of the precious liquid. As far as I can determine, all non-Roman cities into which water was brought from afar had, without exception and until recently, one thing in common: the water that the aqueduct brought across city lines was absorbed by urban soil. Sewers that channeled piped water (as opposed to rain) remained the exception and, where they did exist, were luxuries, not the rule. Even in Rome most of the waters from public fountains flowed into the Tiber over the street pavement along the curbstone. The *cloaca maxima* served some four centuries to dry out the marshes between the Roman hills. Only in imperial times did it become a channel for black waters and was somewhat later covered over.[24] The idea that we now take for granted, that water piped into the city must leave the city by its sewers is very modern; it did not become a guideline for urban design until a time when most cities had railroad stations and their streets began to be lighted by gas.

The modern idea of a "stuff" that follows its destined path, streaming forever back to its source, remained foreign even to Renaissance thought. Harvey's concept of "circulation" represents a profound break with the past. The newness of the idea of circulation is perhaps as crucial for the transformation of the imagination as was Kepler's decision to replace the translucent spheres carrying a luminous planet (in which

---

24. While water supply *systems* in Rome and in provincial cities have been well understood (thanks, in large part, to **Sextus Frontinus**), sewer systems have not been accorded similar attention. For bibliography consult **Pauly** (I, 104-106) and the article on "Beduerfnisanstalt," ibid. Classical texts on sewers are collected in **Martin** (209–11). They show that very special places in Roman and Hellenistic towns had sewers but give us no idea whatsoever as to the percentage of the total urban population served by them.

Copernicus still believed) with the new elliptical orbits traveled by rocky globes. Circulation is as new and as fundamental an idea as gravitation, preservation of energy, evolution, or sexuality. But neither the radical newness of the idea of circulating "stuff" nor its impact on the constitution of modern space has been studied with the same attention that was given to Kepler's laws or to the ideas of Newton, Helmholz, Darwin, or Freud.

Bodies had always been able to *circle* around a center. The abstract concept of circular motion had lent itself to influential metaphors. The presence of the center "altogether at once" at each point of its circle's periphery had been a symbol for God, soul, and eternity. Time, too, was thought by many schools to pass in circles. The phoenix was the symbol of renewal by fire; Plato described cyclical renewal as a periodic flood. Souls were able to be born and reborn. But the connection between "waters" and what we call circulation had not been made. Before Harvey the "circulation" of a liquid meant what we call "evaporation": the separation of a "spirit" from a "water," for instance the distillation of liquor from wine, or the process or "spiritualization" by which blood was assumed to pass through the septum (we now consider impenetrable) from the left to the right of the brain. The idea of a material that flows forever back to its own source constitutes a major innovation in the perception of water, a transubstantiation of its "stuff."

The first liquid to which "circulation" was ascribed was the blood, and the first man, apparently, to have suggested the idea that blood circulates was Ibn al-Nafiz. He was a physician, born in Baghdad, who died a famous teacher and polymath in Cairo in 1288.[25] Starting from the conviction that

---

25. Ibn al-Nafiz was born in Damascus, became a polymath, teacher and then chief medical officer of Egypt. He "never prescribed a remedy when diet would do, never a composite when a simple would do" (*Encyclopedie de l'Islam* III, 921-22). He died next to his luxurious fountain in Cairo in

H₂O and the Waters of Forgetfulness

the septum which divides the right and left ventricle in the heart is impermeable, he postulated the "small" circulation of the flow of blood through the lungs and to the heart. However, his idea was so totally alien to what the common sense of his times held about the behavior of any "stuff," that it was not once mentioned in the many Arabic commentaries on his work. The idea of circulation in our sense remained beyond the imagination of mid-sixteenth-century Europeans. The two physicians of the sixteenth century who suspected what Harvey "discovered" were both dependent on al-Nafiz who reached them through an Italian who had spent thirty years in Syria: they were Servetus, a Spanish genius and heretic burnt by Calvin, and Realdus Colombus of Padua, where Harvey later studied.[26]

I mention this debt of Servetus and Realdus Columbus[27] to al-Nafiz in order to emphasize how unprepared for "circula

---

1288 at the age of eighty. The literary dependence of fifteenth-century Spanish and Italian physicians on his theory of pulmonary circulation has been recognized only recently. See **Meyerhof** and **Schacht**.

26. Miguel Servet was born in 1509 in Navarra, learned Latin, Greek, and Hebrew and was sent to study law at Toulouse in 1528. He traveled to Italy from 1529 to 1530 as secretary of Charles V's confessor. He wrote *De Trinitatis erroribus*, the first modern denial of the divinity of Christ. After seeking refuge in Geneva, he went to France, where he took the name of Michael Villanovanus. He edited Ptolemy's geography in Lyon, where he was also a pupil of the physician Symphorien Champier. He then later became a disciple of Vesalius in Paris (1536). In a chapter on the Holy Spirit, in his treatise *Christianismi Restitutio*, he describes pulmonary circulation. See Ullmann, 176: "fit autem communicatio haec non per parietem cordis medium, ut vulgo creditur, sed magno artificio a dextro cordis ventriculo, longos per pulmones ductu, agitatur sanguis subtilis: a pulmonibus preparatur, flavus efficitur et vena arteriosa in arteriam venosam transfunditur."

27. Matteo Realdo Colombo, 1516(?)–1559, studied under Vesalius in Padua and succeeded him as teacher after the master's departure in 1543. Colombo turned bitterly against his teacher in *De re anatomica*, where he paraphrases al-Nafiz. He died as chief surgeon to Julius III. For informa-

tion" the early seventeenth century was when William Harvey studied medicine in Padua in 1603. Even when the term "circulation" was used in medicine (as for instance by Andreas Cessalpinus) to describe the movements of the blood (and I have found no evidence that it was used in relation to other liquids), it meant a slow and irregular meandering back and forth. In 1616 Harvey began to lecture on the motions of the heart; by 1628 he formally stated his ideas on the double circulation of the blood; and by the end of the century medical science had generally accepted the idea. It was well into the eighteenth century before Harvey's idea was broadly applied in medical practice. In 1750 Dr. Johannes Pelargius Storch, the author of an eight-volume authoritative gynecological guide for practitioners, could still not accept the general validity of Harvey's discovery. He agreed that blood might flow through the bodies of Englishmen and wash out waste material, but in his own patients from lower Saxony the blood was still receding and clumping and ebbing within the body and had to be drawn out through venesection at the right spot. Storch seems to have resisted the **profound redefinition of the body that the circulation of the blood would have demanded of him**. To accommodate circulation, the quivering and symbol-laden flesh of tradition must be recast as a functional system of filters, conduits, valves, and pumps. He seems to have been unable to conceive of the continuity of space below the skin of his patients required by such a system.

At the beginning of the eighteenth century—with the exception of France where ideas had already begun to "circulate"—the term as used in medicine meant the same as that used by botanists to speak of the flow of sap. But then, quite

---

tion, see E. A. Coppola. The discovery of pulmonary circulation is described in *Bulletin of the History of Medicine* 31 (1957):44–77.

H₂O and the Waters of Forgetfulness

suddenly, around 1750 wealth and money begin to "circulate" and are spoken of as though they were liquids.[28] Society comes to be imagined as a system of conduits. "Liquidity" is a dominant metaphor after the French revolution; ideas, news-

28. "Circulation": in classical Latin a *circulator* is an itinerant performer or vendor who gathers impromptu groups around him (*Oxford Latin Dictionary*). Of the several dozen verbs that are formed with 'circum-' all but two are transitive. The two exceptions are 'circumvolo' (applied to persons, 'to flit around' or 'hover') and 'circumsto' ('to be around'). Whatever was done in a circle was conceived as a wheel turning around a center. The travel of barristers and judges around their 'circuits' was expressed with the intransitive 'to circulate' only later. (*OED* gives three dates, 1664, 1672, 1777—all posterior to Harvey—for 'to put into circulation'.) After blood and ideas, water began to circulate; electricity 'circulated' after 1865,

Poulet has analyzed the use of the circle as a metaphor from the Renaissance to modern times. The possibility of conceiving of a "circulation" that does not presuppose a center is opened up only after Giordano Bruno (*De la causa, principio et uno*, in *Opere italiane*, ed. Paul Lagarde, 1888, 278): "Let us state that the universe is all a center, and that this center is everywhere, and the circumference is nowhere." **Mahnke** and, more recently, **Nicholson** have both shown that the significance given to circular movement has changed with the loss of a sense for the "center" in which, according to the earlier Christian interpretation, every part of the circumference was virtually present.

Ferdinand **Brunot** (1966) notes in ch. 7 of his history of the French language that the word "circulation" was not created in the eighteenth century; it was already used in the seventeenth century for the circulation of the blood, the stars, and even ideas. It was also applied to money before the economists used it. (Boulainvilleres 1727; Savary, *Le parfait Négotiant*, 1736, does not use it.) By the middle of the century it is everywhere. (Montesquieu, *Lettres persanes*, 117, says "the more 'circulation' the more wealth" and *L'Esprit des lois*: "multiply wealth by increasing 'circulation'"; Rousseau, **Turgot**: "this useful and fecund circulation that enlivens all society's labors" and "a 'circulation of labor' as one speaks of the circulation of money"—1766.)

According to **Teich**, the "circulation of matter," first conceived by Georg Ernst Stahl, further developed by Priestly and finally formulated by Lavoisier, is one of the very few *major* scientific generalizations on which the twentieth century still builds. It can be compared in importance to the theories of gravitation, conservation of energy, and evolution, but it has not attracted comparable attention from historians of science.

papers, information, gossip, and—after 1880—traffic, air, and power "circulate."[29]

By the mid-nineteenth century some British architects begin to speak of the inner city using the same metaphor.[30] In

---

It would be a mistake to consider these men simply as epigones of classical atomists. To give substance to the preconceived theme of circulation Stahl (1660–1734) invented a primary, common constituent of metals, minerals, plants, and animals that was "inflammable fatty earth" (Brand-und-Fett-Prinzipium) given off in "combustion." He asserted that it was an elusive substance which neither then nor thereafter has been identified. "Phlogiston" ("das erste, eigentliche gruendliche brennliche Wesen"), was the "principal, specific, fundamental essence of all combustion." He described it as both the matter of fire and of its movement. He made phlogiston responsible for the color and odor of substances. He considered chemical change as a circulation of phlogiston. Stahl's phlogiston theory maintained a firm hold on the minds of most mid-century chemists for a good thirty years after his death.

Then in 1767 Joseph Priestly, conducting experiments with "fixed airs" in the Leeds brewery, successfully separated qualitatively different "gases." One of the "airs" he described in 1774 was remarkable because it alone was "respirable" by animals and could be "restored" by growing plants. He called it "dephlogisticated air." In a couple of years Lavoisier began to call it "gas of oxygen." Combustion could now be described as the "combination" of a substance with oxygen—precisely the inverse of the "loss of phlogiston" postulated by Stahl. Eighteenth-century chemistry thus became the science of the circulation of matter that is neither created nor lost.

29. **Dockes** analyses the relationship of sixteenth- through seventeenth-century economic concepts to spatial representations that they imply. He finds that economic policymakers of the seventeenth and eighteenth centuries treat "space" as a physical reality. The movement of capital, work force, raw materials, and products from one space to another is always spoken of as a flow across borders that fuzzily separate nonuniform domains. No doubt, mercantilists expected from the state some homogeneization of these spaces, and liberals resented any such centralizing efforts on the part of governments. However, for both schools a realtionship to social space as an extra-economic reality was implicit in all their analyses.

30. Pierre Patte was an architect and a self-made man. For a time he was draftsman for the illustrations of the encyclopedia of Diderot and got

H_2O and the Waters of Forgetfulness

1842 Sir Edwin Chadwick, former literary assistant to Jeremy Bentham and afterwards member of the royal commission on the poor laws, presented a report on the sanitary conditions of the laboring population of Great Britain. Lewis Mumford has called it the "classic summary of paleo-technic horrors." In this report Chadwick imagined the new city as a social body through which water must incessantly circulate, leaving it again as dirty sewage. Without interruption water ought to "circulate" through the city to wash it of its sweat and excrement and wastes. The brisker this flow, the fewer stagnant pockets that breed congenital pestilence there are and the healthier the city will be. Unless water constantly circulates through the city, pumped in and channeled out, the interior space imagined by Chadwick can only stagnate and rot.

Chadwick's papers were published under the title "The Health of Nations" during the centenary commemoration for Adam Smith. Like the individual human body and the social body, the city was now also described as a network of pipes. The brisker the flow, the greater the wealth, the health, and the hygiene of the city. Just as Harvey redefined the body by postulating the circulation of the blood, so Chadwick redefined the city by "discovering" its need to be constantly washed.

The Dirt of Cities

When Aristotle drew up his rules for the siting of a city, he wanted the streets to be open to sunlight and to prevailing

---

involved in an interminable legal struggle with his employer. He saved the Paris Pantheon from crumbling. His fancy plans for providing Paris with water and sewerage ("Sur la distribution vicieuse des villes," 28ff. in *Mémoire sur les objets les plus importants de l'architecture*, Paris, 1769.), according to J. Rykwert is often forgotten. For others who made recommendations for the improvement of Paris water, see **Roche**, 385–391. Nonetheless, there seems to be no doubt that the ideology of circulating waters was formulated effectively for the first time by Chadwick. For his

winds. Complaints that cities can become dirty places go back to antiquity. In Rome special magistrates sat under their umbrellas in a corner of the Forum to adjudicate complaints from pedestrians soiled by the contents of chamberpots. Throughout classical antiquity, beginning with the palace at Knossos (1500 B.C.), the dwellings of the wealthy occasionally had a special room for bodily relief. In Rome, wealthy households owned a special slave to empty the night-chairs. Most homes had no designated place for bodily relief. Like the sewers beneath the Athenian agora, the sewers beneath the imperial fora and pay seats in marble latrines were restricted to city areas covered with marble. In popular two-story dwellings, Roman ordinances required a hole at the bottom of the staircase. Otherwise, the street was assumed to be the proper place for such disposals. Medieval cities were cleaned by pigs. There survive dozens of ordinances which regulate the right of burghers to own them and feed them on public waste. In Spain and the Islamic areas, ravens, kites, and even vultures were protected as sacred scavengers. These customs did not change significantly during the baroque period. Only during the last years of Louis XIV's reign was an ordinance passed that made the removal of fecal materials from the corridors of the palace in Versailles a weekly procedure. Underneath the windows of the ministry of finance, pigs were slaughtered for decades, and their encrusted blood caked the palace walls. Tanneries were operated within the city, even though their smell in the valley of Ghinnom had become the symbol for hell (gehenna) in old Jerusalem. A survey carried out in Madrid in 1772 disclosed that the royal

---

biography see **Dickinson** and **Beguin**. The one major monograph by a first-rate historian on the whole history of water in western Europe is by **Robinson**. More than half of the book is concerned with the nineteenth century.

palace did not contain a single privy. These millennial city conditions prevailed in London when Harvey announced his discovery of the circulation of the blood.[31] Only after the great London fire of 1660 and after Harvey's death were "laystalls" set up on London street-crossings for the disposal of waste, and an honorary scavenger was appointed for each ward to supervise the rakers—men and women willing to pay for the privilege of sweeping the streets so that they could sell the refuse for a profit. In 1817 the powers of these scavengers and rakers were codified in the London Metropolitan Paving Act, which remained the statute until 1855. By this time the houses of the well-to-do in London usually contained one privy, from which the night-soil was removed several times each week. But for the larger part of London, the collection of night soil from the streets remained sporadic. In the late nineteenth century it was felt to interfere with rush hours. It was not until 1891 that the London County Council prescribed that privy cleaning had to be restricted in summertime to the hours between 4 A.M. and 10 A.M. Quite obviously, throughout history cities have been smelly places.

The Aura of Cities

Nevertheless, the perception of the city as a place that must be constantly washed is of recent origin. It appears at the time of the Enlightenment. The reason most often given for this constant toilette is not the visually offensive features of waste or the residues that make people slip on the street but bad odors and their dangers. The city is suddenly perceived as an evil-smelling space. For the first time in history, the

---

31. For a general introduction to the history of sanitation, see **Rawlinson** and also **Kennard**. **Gay** is anecdotal and not documented. For London in late medieval times there are many facts on street cleaning and the technique of cesspool construction in **Sabine 1934, 1937**. For the hygienic conditions of Paris streets, **Labande** is full of details, **Gaiffe** dated and amusing.

utopia of the odorless city appears. This new aversion to a traditional characteristic of city space seems due much less to its more intensive saturation with odors than to a transformation in olfactory perception.

The history of sense perception is not entirely new. Linguists have dealt with the changing semantics of colors, art historians with the style in which different epochs see. But only recently have some historians begun to pay closer attention to the evolution of the sense of smell. It was Robert Mandrou who, in 1961, first insisted on the primacy of touch, hearing, and smell in premodern European cultures. Complex nonvisual sense perceptions gave way only slowly to the enlightened predominance of the eye that we take for granted when we "describe" a person or place. When Ronsard or Rabelais touched the lips of their love, they claimed to derive their pleasure from taste and smell, which could only be hinted at. Even the eighteenth-century writer does not yet describe the loved body; at best the publisher inserts into the text an etching that illustrates the scene, an etching which, during the early part of the century, effectively hides whatever is individual, personal, "touching" in the scene the author describes. But while it is easy to follow historically the ability of poets and novelists to perceive and then paint the flesh and landscape in their uniqueness, it is much more difficult to make statements about the perception of odors in the past. To write well about this past perception of odors would be a supreme achievement for a historian because the odors leave no objective trace against which their perception can be measured. When the historian describes how the past has smelled he is dependent on his source to know what was there and how it was perceived. The case is the same whether he deals with odors perceived by lovers or those that help physicians recognize the state of

H_2O and the Waters of Forgetfulness

the ill or those with which devils or saints fill the spaces within which they dwell.[32]

I still remember the traditional smell of cities. For two decades I spent much of my time in city slums between Rio de Janeiro and Lima, Karachi and Benares. It took me a long time to overcome my inbred revulsion to the odor of shit and stale urine which, with slight national variations, makes all unsewered industrial shantytowns smell alike. This smell is the characteristic for the early stage of industry; it is the stench of dwelling space that has begun to decay because it is threatened by imminent incorporation into the hygienic system of modern cities. It is distinct from the local atmosphere of a still vernacular town. A vernacular atmosphere is integral to dwelling space; according to traditional medicine, people waste away if they are sickened and repelled by the aura of a new place in which they are forced to live. Sensitivity to an aura and tolerance for it are requisites to enjoy being a guest. Many people today have lost the ability to imagine the geographic variety that once could be perceived through the nose. Because increasingly the whole world has

---

32. Stench that kills not one but several persons on the spot is not new to the mid-eighteenth century. There are many previous reports of sinners killed on the spot by experiencing the devil's stench. What is new is the connection between the stench of decaying bodies and this physical effect. See **Foizil** and **Ariès**. All through the Middle Ages the sense of smell opened the gates of heaven and hell. Reports on the "odor of sanctity" perceived year after year by thousands of visitors to the grave of a saint are quite common. **Deonna** documents several hundred instances. **Lohmeyer, Nestle,** and **Ziegler** relate this experience to biblical texts. During the twelfth century the smell particular to the remains of saints was taken as evidence for the authenticity of such relics. There can hardly be any doubt about the widespread sharing of this experience. The perception of space and its characteristics by means of the sense of smell was taken for granted by the poets of the time. See **Hahn** ("Duftraum") and **Ruberg,** 89ff. A beautiful introduction to the "meaning" given smells is in **Ohly.** See also **Ladendorf.**

come to smell alike: gasoline, detergents, plumbing, and junk foods coalesce into the catholic smog of our age. Where this smog mingles with the decay of vernacular atmosphere, as along the Rimac which carries Lima's sewage into the Pacific I learned to recognize the smell of development. It is there that I became sensitive to the difference between industrial pollution and the dense atmosphere of Paris between Louis XIV and Louis XVI. To describe it I shall draw heavily on Corbin.

The Smell of the Dead

People then not only relieved themselves as a matter of course against the wall of any dwelling or church; the stench of shallow graves was evidence that the dead were present within its walls. This thick aura was taken so much for granted that it is rarely mentioned in contemporary sources. Universal olfactory nonchalance came to an end when a small number of citizens lost their tolerance for the smell of corpses. Since the Middle Ages, the corpses of clergy and benefactors had been entombed near the altar, and the procedures of opening and sealing these sarcophagi within the church had not changed over the centuries. Yet at the beginning of the eighteenth century, their miasma became objectionable. In 1737 the French parliament appointed a commission to study the danger that burial inside churches presented to public health. The presence of the dead was suddenly perceived as a physical danger to the living. Philosophical arguments were concocted to prove that the burial within churches was contrary to nature. An Abbé Charles Gabriel Porée, Fénelon's librarian, from Lyons argued in a book which went through several editions that, from a juridical point of view, the dead had a right to rest outside the walls. In his monumental history of attitudes toward death in the West since the Middle Ages, Philippe Ariès has shown that this new squeamishness in the presence of corpses was due to an equally new unwillingness to face death. Henceforth, the living refused to share their space with the dead. They demanded a special apartheid between live bodies and corpses at just the time when the innards of the live human body were beginning to be visualized as a machine whose elements were "prepared" for inspection on the dissecting table. Like the organs, the dead became more visible and less awesome; they also became increasingly more disgusting and physically dangerous for the living. Philosophical and juridical arguments calling for their exclusion from dwelling space went hand-in-hand with reported evidence of the deadly threat of their miasma. Corbin lists several instances of mass death among the members of a church congregation that occurred at the very moment when, during a funeral ceremony, miasma escaped from an opened grave. Burials within churches thereafter became rare—increasingly a privilege of bishops, heroes, and their like. The cemetaries were moved out of the cities. Though in 1760 the Cimetière des Innocents was still used for parties in the afternoon and for illicit love at night, it had been closed in 1780 by request of neighbors precisely because they objected to emanations from decomposing bodies. Yet even if the presence of the dead within the city was resented by rich and poor alike at the end of the ancien régime, it required almost two centuries to educate the lower classes to feel nausea from the odor of shit.

Utopia of an Odorless City

Both living and dead bodies have an aura. This aura takes up space and gives the body a presence beyond the confines of its skin. It mingles with the auras of other people; without losing its own personality, it blends into the atmosphere of a particular space. Odor is a trace that dwelling leaves on the environment. As fleeting as each person's aura might be, the atmosphere of a given space has its own kind of permanence, comparable to the building style characteristic of a neighborhood. This aura, when sensed by the nose, reveals the non-dimensional properties of a given space; just as the eyes perceive height and depth and the feet measure distance, the nose perceives the quality of an interior.[33]

During the eighteenth century it became intolerable to let the dead contribute their aura to the city. The dead were either excluded from the city or their bodies were encased in airtight monuments celebrating hygienic disposal, for which

33. Giving off a smell is as much a part of a personality as casting a shadow, producing a mirror image, or leaving traces on the ground. In all of these "aura" becomes perceptible. People recognize one another by smelling out where they come from: "The Scots folks have an excellent nose to smell their countryfolk" (*OED*, 1756). One first relies on smell to discriminate among individuals: "What a man cannot smell out, a man may spy into" (*King Lear* 1, v, 23:1605). But "you can easily smell a rat," except that "where all stink, no one is smelled." The Latin proverb "mulier tum bene olet ubi nihil olet" quoted by Plautus has been variously translated: in 1529 as "A woman ever smelleth best, whan she smelleth of nothing" and in 1621 by Burton as "Then a woman smelleth best, when she hath no perfume at all." During this period "perfume" had changed its meaning. It had come into English as "odor", given off by incense or other burning substances and, by the time of Burton, had come to mean "scent." (See also **Tilly**, nos. S558 and R31, and F. **Wilson**, under "smell.")

During the second decade of the nineteenth century, the loss of "aura" becomes a major new motif in literature. It can be readily traced by following the influence of A. V. Chamisso's *Peter Schlemihl*, who sells his shadow to the devil in exchange for wealth. The loss or sale of one's "soul" was a well-known motif at the time, but by retelling the folktale and insisting on the loss of something visible and observable, Chamisso created a veritable school. In 1815 E. T. A. Hoffmann told the story of a young man whose mirror image was taken from him by a whore and an eerie physician. W. Hauff's hero of 1828 exchanges his heart for a stone counterfeit to save himself from bankruptcy. By the end of the century heroes have sold "sleep," "appetite," "name," "youth," and "memories" (for details, see **Ludwig** 1920 and 1921).

The shadow had always been part of the full personality (Bächtol 9,

H_2O and the Waters of Forgetfulness

Père Lachaise became the symbol in Paris. In the process of their removal, the dead were also transmogrified into the "remains of people who have been," subjects for modern history —but no more of myth. Disallowing them shared space with the living, their "existence" became a mere fiction and their relics became disposable remains. In this process western society has become the first to do without its dead.

The nineteenth century created a much more difficult task for deodorants. After removing the dead, a major effort was undertaken to deodorize the living by divesting them of their aura. This effort to deodorize utopian city space should be seen as one aspect of the architectural effort to "clear" city space for the construction of a modern capital. It can be interpreted as the repression of smelly persons who unite their separate auras to create a smelly crowd of commonfolk. Their "common" aura must be dissolved to make space for a new city through which clearly delineated individuals can circulate with unlimited freedom. For the nose a city without aura is literally a "Nowhere," a *u-topia*.

Nachtrag 126–42). Only when a Greek becomes luminous himself, in the presence of Zeus or when an Iranian becomes a saint, does he lose his shadow. According to Irish stories, if a person's shadow is pierced, he dies (**Stith-Thompson** D 2061.2.2.1). Among the Jews a ghost is recognizable by its lack of a shadow (ibid., G 302.4.4), just as it is said to leave no footprints (ibid., E421.2). **The exchange in which the student of alchemy leaves his shadow to his master the devil as an honorarium is a motif that appears only in the eighteenth century.** The shadow remains secondary in fairy tales and folk literature (Franz 1983). In fairy tales everybody is always everybody's shadow (24, 31). The idea of the shadow (or, for that matter, of the mirror image or act of memory) as a saleable commodity is a new and important motif that appears with possessive individualism in Chamisso. It fits into the period during which people's "smell," their aura and their "moral economy" (E. P. Thompson) were taken from them.

Ultimately the drugstore became the symbol of the industrialized aura; it is the supermarket of mass-produced glamour and scents for a deodorized population. People who obsessively scrub away their auras can pick

The clearing of city space coincides with a new stage of the professionalization of architects. Their profession had formerly been in charge of building palaces, squares, fountains, city walls, and perhaps bridges or channels. They were now empowered to condemn dwelling space and transform it into garages for people. Observing the course of Peruvian settlement thirty years ago, John Turner has described what happens when dwelling *by* people is transformed into housing *for* people. Housing is changed from an activity into a commodity. This transformation requires making dwelling activities impossible, so that persons become domesticated docile residents within shelters which they rent or buy. Each now needs a street address with a house number (and, in some cases, an apartment number too). People have lost the aura that allowed their whereabouts to be sniffed out in the old days. When the idea of the new city, made up of residents, began to register in the minds of the leaders of the Enlightenment, everything that smacked of quality in space came to be objectionable. Space had to be stripped of its aura once aura had been identified with stench. Unlike the architect who constructed a palace to suit the aura of his wealthy patron, the new architect constructed shelter for a yet unidentified resident who was supposed to be without odor.

Miasma as Turned Out Gas

The most outspoken members of that elite which, around 1750, called attention to the dangers of city miasma were not architects, however, but students of pneumatics—the discipline that specialized in research on breath, spirits, or airs. The term "gas" had been coined, but it was not then in use. It would have been of little use for this research because, for

---

and choose a better one there. Musil (v. 7, p. 895) has created a prophetic image: "Schlemil's guilt is his bourgeois nature, his refusal to admit his loss of his shadow, his incapability of creating genius from it."

the alchemist J. B. van Helmont who first wrote it down, it was a translation (in Dutch phonetics) for the "chaos" that Paracelsus understood as the common term for the "elements" insofar as these are dwelling spaces of elementary spirits. It is not easy for a twentieth-century person to imagine the helplessness of a scientist in the first half of the eighteenth century when he tries to analyze what we currently know as the "gaseous state." The instruments for the study of volatile substances were still rudimentary at that time. The researcher had to rely mostly on his nose.

Combustion was not yet understood to be a process of oxidation but was thought to be the release of "phlogiston" from the burning body into the air. This phlogiston that escaped from the flame into space was—partly for leftover theological reasons—considered to be a "fatty kind of earth." Within this context of helpless uncertainties, the fantasy that went into the study of "odors" is understandable. The sense of smell was the only means for identifying the city's exhalations.[34]

34. "Gas": when speaking of the four elements, Paracelsus used the term "chaos" to designate them as the appropriate "space" for whatever dwells in them naturally: ". . . the little mountain men have the earth as their chaos; now for them it is only an air and no earth such as for us. From that it follows that they see through earth as we see through air. . . . For the undines water is chaos; now for them the water does not hinder the sun; just as we have the sunlight through the air, so they have it through the water . . . of volcanic masses also through their fire. . . ." (*Paracelsus*, ed. **Blaser**, 20). Elsewhere, he explains that "earth is no more than the sole chaos of these little mountain men . . . since they go through it, through chains or stone, like a ghost . . . as little as we are hindered in going through the air." (Paracels, *Werke*, J. Huser: Basel, 1590, 54.)

The Latin pronunciation of "chaos" gave the "ch" combination ("x" in Greek) the same guttural sound as "ch" in German and "g" in Dutch. J. B. van Helmont chose to respell Paracelsus's term in conformity with Dutch orthography. The result was a term whose original connection with alchemical language and concepts was gradually forgotten. As a result, we are apt to misunderstand what is meant by the work in these earlier historical contexts. The act of obtaining this "spirit" or gas from matter was for van Helmont a "spiritualization" of matter. The conception of gas as a form of matter thus contributed to the idea of the circula-

Ivan Illich

The osmologists (students of odors) collected "airs" and smelly materials in tightly corked bottles and compared notes by opening them at a later time as though they were dealing with vintage wines. A dozen treatises focusing on the odors of Paris were published during the second part of the eighteenth century. They deal with the classification of odors that coincide with the stages of decomposition of a carcass, with the seven smelly points that lie between the top of the head and the interstices between the toes, with the distinction between the healthy strong smell of dung and human excrements and the putrid and dangerous emanations of decay. One treatise even estimates the weight of per capita exudations of city dwellers and the effect of this pollution when it is deposited in the city's vicinity. In almost every book on this subject, the author includes a bitter complaint about the insensitivity of the general public to the dangers of bad airs he has discovered and described.

By the end of the eighteenth century, this avant-garde of deodorant ideologues is causing social attitudes toward body

---

tion of matter. Long confined to technical use (Kruenitz 1779), the word "gas," according to Wieland (T. Merkur 1, 75) was still unknown in Germany in 1784. That very year, Minckelaars recommended gaslighting, but the German encyclopedist Adelung opposed the use of that word: "The word is barbaric and obscure ... our naturalists need to find a more suitable word, one having less of the character of alchemy about it" (after Truebner WB 2, 425 cit).

The first French use, is given in the *Thesaurus de la Langue Française*: Benjamin Constant, *Journaux*, 1804, 20: "The old chemists called spirits aeriform or gas, since they had not yet discovered the art of collecting or fixing *spiritus silvestres*, wild spirits, and they refused to have anything to do with them." Brillat-Savarin (*Physique du Gout*, 1825, 42) saw in the new concept a threat to his theory of high taste: "The sapid body is only esteemed for its juice and not for the odoriferous gas that emanates from it" (from the same source). Goethe, *Faust II*, 4, v, 10084: "Hell swelled up with stench and sulfuric acid, which gave off a gas. It went on to become immense." This painful search for the idea of "gas" reflects the simultaneous attempts to materialize smells.

wastes to change. The king's audience *en selle*, for those who were specially privileged, had ceased two generations earlier. Toward the middle of the century shitting, for the first time in history, became a sex-specific activity: separate latrines for men and women are set up but only on special occasions. At the end of the century, Marie Antoinette has a door installed to make her own defecation private. The act turns into an intimate function. After the process has been thus removed from sight—if not from the nose—its product is also shoved out of reach. Defecation and urination are hidden in the closet. During the Napoleonic wars, English and French upper-class latrines move in two different directions. In France the ornate stools, that in the eighteenth century were part of boudoire furniture, are moved into special closets; they continue to be regularly cleaned by servants, and as a result their presence in the upper-class household is now less easily perceived. Starting in the late eighteenth century, the English upper classes adopted the WC. It was usually in a closed cupboard and connected by unventilated pipe to a cesspool in the cellar. The unintended feedback from this progress in hygiene was the saturation of English townhouses with a new type of gas resulting from more advanced stages of decomposition. While the English got used to it as the appropriate aura of elites, foreign visitors during the entire century commented on this peculiar phenomenon without, however, recognizing its technical source.

Stooling and Privacy

Not only excrement but the body itself, it was discovered, emanates bad odors. Underwear that up to this time had served to keep one warm or attractive began to be connected with the elimination of sweat. The upper classes began to use and wash it more frequently, and in France the bidet came into fashion. Bed sheets and their regular laundering acquired a new importance, and to sleep in one's own bed between sheets was charged with moral and medical significance. For young men, heavy blankets were proscribed because they accumulate body aura and lead to wet dreams. In 1780 the Hôtel Dieu ruled that each person who recovered in the hospital would be placed henceforth in a separate bed, but this hygienic ideal did not become practice in most hospitals until after the Congress of Vienna. On 15 November 1793, the revolutionary convention solemnly declared each man's right to his own bed as part of the rights of man. Each citizen has the right to be surrounded by a buffer zone that protects him from the aura of others, while keeping his own to himself. The private bed, stool, and grave became requisites of a citizen's dignity. To insure at least one of these to everyone and to spare the poor citizen at least the horror of burial in a mass grave, charities popped up during the turn of the century.

Parallel with the privatization of body relief and the attempt to retrench people's auras, reducing each to an odorless point in the new civic space, the toilette of the whole city was undertaken. The first places that had attracted the attention of the reformers were prisons and bedlams, with their knee-deep filth that could be smelled even from a distance. Those who waited for judgment, flogging, or transportation, those who were locked up for a term—all criminals, waifs and the mad—were thrown together, and the high mortality rate by midcentury was attributed to the intolerable atmosphere of the place. The ventilator had just been invented for onboard use, and the first to be installed on shore was used to give a whiff of fresh air to those sections of the prison where the innocent inmates were being kept. Air for prisoners seemed to be a difficult problem for the administration, so several cities between the Alps and the Netherlands adopted the idea of a Dr. Berne which allowed them to combine the removal of excrement from the street and the ventilation of prisoners. The new machine built for the purpose was a cart drawn by chained men to which women were attached by lighter restraints to allow them free movement over the pavement from which they picked up the night soil, dead animals, and other refuse. This cart could be deployed especially in those sections of the city which, by analogy with the human body, were recognized as the city's smelly points.

Osmologists Discover the Smell of Race and of Class

Smell now began to become class-specific. Medical students observed that the poor are those who smell with particular intensity and, in addition, do not notice their own smell. Colonial officers and missionaries brought home reports that savages smelled differently from Europeans. Samojeds, Negroes, and Hottentots could each be recognized by their racial smell, which changes neither with diet nor with more careful washing. Social advance came to be identified with increased cleanliness. One could move into the better classes only by getting rid of body smell and making sure that no odor attached to one's home. Water became a detergent of smell.

Throughout history the degree of contact of water with human skin has varied widely from culture to culture. Until the 1930s, in many areas of France and England most infants' skins were carefully shielded from water and wiped only with a handkerchief moistened with their mothers' spittle. In many areas, more than half of the population had never taken a bath at the time they died. They were washed when they were born and again after death. In other cultures weekly bathing, steaming, sweating, and scraping of grime from the skin was a must.

Only during the nineteenth century soap came to be associated with body laundry. Previously, hard soap was a precious cosmetic, and a home-made paste of potash was used on fabrics only. Soap is the first industrial product to create its own demand and engage the school system as a publicity agent. Development, right into the late twentieth century, has remained associated with water and soap.[35]

Slowly—and in different decades for different income levels—education has shaped the new sense for cleanly individualism. The new individual feels compelled to live in a space without qualities and expects everyone else to stay within the bounds of his or her own skin. He learns to be ashamed when his aura is noticed. He is embarrassed at the thought that his origins could be smelled out, and he is sickened by others if they smell. Shame at being smelled, embarrassment at coming from a smelly environment, and a new proneness

35. The removal of fatty substances from the skin by means of soap seems to have been unknown to the Romans. In Europe up until the seventeenth century, it remained a procedure executed upon the advice of a physician. Pliny (*Hist. Nat.* 28, 191) reports that Germanic men used it to bleach their hair red to frighten outsiders in battle. During the Middle Ages, soap was in widespread use for laundry. The three traditional crafts that depend on potash and soda were usually connected: glass making, dyeing, and soap boiling. Of course soap boiling could go on only where fats not consumed as food were available. Hard cakes of olive-based scented soap, imported from southern Europe, have been a luxury since the fourteenth century. During the eighteenth century, whale hunting and the demand for soap reinforced one another. Soap became cheaper and soap cakes more common. Only after 1780 was the process of saponification understood scientifically. This made it possible to calculate the quantities of needed ingredients more precisely and to produce soap on a large scale. Within another forty years the application of the Leblanc process to industrial soapmaking led to the first public recognition of the environmental dangers represented by the chemical industry. Large amounts of hydrogen chloride gas were produced and dispersed in the air through tall chimneys. Widespread devastation of vegetation and even forest resources ensued. In 1828 the first lawsuit was initiated against the factory with which Mr. Gamble was associated in order to obtain protection against environmental damage. By the end of the century hydrogen chloride found industrial applications not only in bleaching but also in the clorination of drinking water.

to be offended by smell—all taken together place the citizen in a new kind of space.

The Educated Nose: Shame and Embarrassment

Three convergent attitudes result from the social repression of smell.[36] One is shame. According to Norbert Elias civilizing shame can be understood as an habitual fear of the humiliating put-down due to one's own uncleanliness. Because the person who is thus shamed has no excuse for imposing his smelly presence on others, his anger turns within; it is by now too late to wash, so he blushes. The second is em-

---

36. When discussing "smell" in English, one must remember that there are many languages—and I am confining myself to the Indo-Germanic family—in which connotations that are only implicit in English are quite explicit. In English my sense of smell allows me to engage in the activity of smelling the smell of a rose that smells. In Italian the two verbs are distinct: "*sento* la rosa che *odora*." In Serbo-Croatian the nouns are from different roots: the rose emanates a sweet *miris* or *vonj*. But that which I experience is *njuh* or *osjet*. The majority of the words for "fragrant" are derived from words of smell, sometimes with a prefix that says "well" or "sweet," but much more often based on a restriction of "smell" to "good smell." For "bad smell" there are clearly two distinct formations: those which are derived from "fragrant" with a prefix meaning "bad," and those which explicitly mean "stink," "rotten," "decay." Words that speak of "odors" and "smell" are apt to bear a strong emotional value that is felt to a lesser degree in words for "taste" and hardly at all in those referring to the other senses (see **Buck**, ch. 15, nos. 15.21-26). We lack an independent classification of smells analogous to that of taste (sweet, bitter, salty) or sight (colors, shapes), as Aristotle has already noted in *De Anima* (2.9). We cannot indicate to each other what we smell except by analogy with another sense or by indicating the object that we smell. We are limited to making the distinction between "good" and "bad" that marks the extremes of a continuum. We are speechless about the central part of the spectrum, the continuous perception of the aura within which we move. This continued perception of space through the nose—nondimensional, complex, and profoundly orienting—must have been experienced neither as a stench nor as a fragrance. Old German has three times as many words for fragrance as modern German. It is my thesis that the increasing monopoly of Cartesian dimensionality over the sensual perception of space weakened or extinguished the sense of aura.

barrassment. Embarrassment is distinct from shame; it is the sting of one's self-consciousness about a stained, blemished, or smudged background that is provoked by noticing the smell of another member of one's own milieu as he too ventures out into the odorless space of the city. Third, shame combines with the fear of embarrassment and a new olfactory delicacy develops; just as each individual keeps tightly wrapped up in himself, his squeamishness, now civilized, keeps him out of other people's private spheres. Each becomes a skunk for the other.

Perfume and the Domestication of Aura

In this new space without quality peopled by uptight spheres of odorless privacy, perfume acquires a new meaning. When the gentleman comes home and takes off his overcoat, he fancies entering a domestic sphere filled with his woman's individual fragrance. Perfumes become sex-workers. Perfume now artificially provides secondary sexual characteristics to the new "human" body stripped of its aura. Like so many other characteristics—for instance, work, health, education—smell, too, henceforth is conceived as an abstract quality that "naturally" is polarized into a male and female type: she smells of violets and roses and he of leather and tobacco.[37]

37. During 1984 I have had several occasions to discuss public issues related to the historicity of smell. At least one person on each of these occasions has called to my attention that, by so doing, I was repressing and sublimating an "event" which took place in prehistoric times. Infallibly the conversation led to Sigmund **Freud**, *Civilization and Its Discontents*, 99: "After primal man had discovered that it lay in his own hands, literally, to improve his lot on earth by working, it cannot have been a matter of indifference to him whether another man worked with or against him.... Even earlier, in his apelike prehistory, man had adopted the habit of forming families, and the members of his family were probably his first helpers. One may suppose that the founding of families was connected with the fact that a moment came when the need for genital satisfaction no longer made its appearance like a guest who drops in suddenly, and, after his departure, is heard of no more for a long time, but instead took up its quarters as a permanent lodger. When this happened,

H_2O and the Waters of Forgetfulness

Not only in meaning but also in chemistry, fragrance is a product of the age. Perfume is old stuff, but the modern history of perfumes as articles of fashion begins when Catherine de Medicis arrived in Paris with a profumer in her train. René de Florence established his boutique on the Pont du Change and soon acquired a large clientele. Using mostly substances known for hundreds of years, he broke with tradition: he specialized in the preparation of personalized prescriptions. Then, under Louis XIV, fragrance began to be dictated by each season's fashion rather than chosen by personal fancy. Each new season of court dictated a new fragrance. Enormous sums were spent to accelerate the succession of waves, each bringing into fashion a different perfume. The substances used were overwhelmingly based on animal products: ambergris, musk, civet, and other excretions from the genitals of rodents. However, for a time under Marie

---

the male acquired a motive for keeping the female, or, speaking more generally, his sexual objects, near him. . . ." See also the footnote to this passage: "The organic periodicity of the sexual process has persisted, it is true, but its effect on psychical sexual excitation has rather been reversed. This change seems most likely to be connected with the diminution of the olfactory stimuli by means of which the menstrual process produced an effect on the male psyche. . . . The taboo on menstruation is derived from this 'organic repression' as a defence against a phase of development that has been surmounted. All other motives are probably of a secondary nature. . . . The diminution of the olfactory stimuli seems itself to be a consequence of man's raising himself from the ground, of his assumption of an upright gait; this made his genitals, which were previously concealed, visible and in need of protection, and so provided feelings of shame in him."

These passages from Freud have strongly influenced many people who have reflected on the history of smell. People in the early twentieth century could not come to terms with the recent vintage of this change. See Jacques **Guillerme** (1977). Many people simply refuse to recognize that strong smell, at least in France, was associated in the eighteenth century with good health. See **Thuillier** (1968, 1977). A fatty surface, particularly in children, was valued as a protection against disease. Warm baths were associated with sensual pleasure and sin. The "polarization" of perfume is a nineteenth-century bourgeois phenomenon.

Antoinette the fashion turned toward lighter oils extracted from vegetable substances; but Napoleon, as an upstart, returned the smell of the court to that of animal glands. Only after the Congress of Vienna did flowers, mostly in toilet waters, come to dominate the "salon." The atmosphere of the romantic bourgeois smelled in extreme contrast to the court. The well-to-do lady began to wash with scented soap and began to enhance her personal flair by sprinkling herself with vegetal fragrances. These are much more volatile and teasing. They are "light" and must be frequently reapplied; they linger in the domestic sphere and are symbolic of conspicuous consumption.[38] Rousseau's Emile is taught that "fragrance never gives as much as it makes you hope for." By the time of Napoleon III, the use of the earlier extracts from sex glands had become a sign of debauchery. By the middle of the nineteenth century the rich are ever so lightly scented, the middle classes are well scrubbed, and the deodorizing of the impoverished majority has become a major goal in the campaigns of educators and for the medical police.

38. On the technique of preparing perfumes in antiquity, see **Forbes (1965)**, which contains comments on texts by Theophrastus, Dioscorides, and Pliny. The treatment of odors in the poetry of antiquity is examined in **Lilja** and **Detienne**. Techniques in the preparation of perfumes changed during the first half of the nineteenth century, but the theory about the sense experience involved in their use did not. See Larousse, vol. 12, article "Parfum" (published in 1876!). "Is the fragrance coming from a body an imperceptible and imponderable gas or rather a dynamic action that hits the olfactory nerve, somewhat in the same way that light acts on the retina of the eye and sound on the sense of hearing? A patient researcher has now shown mathematically that a package of musk left lying in a space thirty meters across for twenty-four hours had lost fifty-seven particles without the slightest diminution in weight. One theorist proposes that fragrances be conceived of as though they were vibrations affecting the nervous system in the manner of colors, and it is quite legitimate to make the supposition that certain bodies emit odor-waves, just as diamonds project light-waves or harps sound-waves. With marvellous speed these *vagues d'odeurs* travel over great distance and probably have nutritive value."

H_2O and the Waters of Forgetfulness

Water is Adopted for the Toilet

The old perfumes had been part of a "toilette" when the word carried no connotations of water. The term "toilette" in the eighteenth century referred to combing, grooming, powdering, applying makeup and perfumed cosmetics, dressing, and then, as a last stage of "toilette," receiving visitors in the boudoire. The "toilette" was hydrophobic, in no way connected with running water. Medical consensus took water to be unhealthy for the skin. If it entered into the toilette, it was to moisten a towel. Where Moors, Jews, or Finns had introduced Europe to the bathhouse, it was used primarily to enhance well-being, not appearances. Frequent cleansing by means of water did not became part of the toilette before the nineteenth century. By the third decade the word came to mean the sponging of a naked body that was always represented as belonging to a woman. From decade to decade, the amount of water used in the procedure increased. The toilette came to mean a tub bath. Neighborhood entrepreneurs began to rent copper basins for the purpose. Then, around 1880, the industrial production of enamel paints replaced expensive copper with iron or zinc vessels and brought the tub within reach of simple families. Later the shower replaced the tub.

The installation of a "bath" room within the apartment fused three formerly different activities: bathing, body cleansing, and dressing for day or night. It became the place where the stool stands and where men shave themselves instead of being barbered. "Toilette" retired behind locked doors. It now involves the flow of tap water to carry soapy suds and excrement to the sewer.

The total bathroom was not invented overnight. When Mlle Dechamps, an opera singer returning from London to Paris around 1750, installed two separate mirror-walled cabinets, one for the faucet and one for the sink, she became the talk of the town. A hundred years later, the bathroom was still a rarity, and those who could afford it put the stool, the faucet, and the dresser in three separate cabinets. With another hundred years, one out of every three to five urban rooms is a toilet.[39] If Pierre Patte, the first architect to design a system (never constructed) of modern sewers for Paris in 1769, had to report back to his patrons on the configuration of a contemporary city, he would have to say that his successors had built it around bathrooms and garages, accomodating the circulation of tap water and of traffic.

It is revealing that the place at which the modern body is integrated into the circulation of city waters is called the "bath" room.[40] This term does not appear in the first edition of the *Oxford English Dictionary*; it is first mentioned in the Supplement (1972) and its initial use traced to 1888. The choice of this term indicates that the identification of nature and the nude which Ingres, Courbet, Corot, and Renoir had painted as taking place in rivers, under waterfalls, or in an oriental *hamam*, was actually performed in the intimacy of the toilette.

---

39. "Toilette" originally meant (fourteenth century) the cloth in which artisans gather their tools; then (sixteenth century) the cloth on which brush and comb are laid out; then (seventeenth century) their use but also the spread of clothes and the act of putting them on.

40. A history of the meaning that was given in western history to bathing has not been written. Inevitably it has been touched upon in all historical treatments of baptism. For the early Christian attitudes to baths see **Jüthner**; for the Middle Ages **Vogüé** (1100–03) gives texts. Bathing for health and for enjoyment disappears from many areas of Europe with the Reformation. **Thorndike** lists, with English commentaries, the older German literature of balneology: bathing for medical reasons. **Wright** is by far the most readable popular account of bathroom fixtures but lacks references. The first part of the book is misleading because the author draws conclusions about supposedly common behavior while forgetting that the contraptions that have survived could be the exception rather than representative discoveries. For the history of the "bathroom" during the later nineteenth century, he is informative and

The Fertile Night Soil of Paris

The use of water for the cleaning of the body and the use of water for the "toilette" of city spaces go hand-in-hand but not at the same pace in all modern nations. Paris never followed the example of London. A 1835 report from L'Institut de France rejected the proposal to adopt the WC and channel excrements into the Seine. The decision was motivated neither by anti-British sentiment nor by concern for the river but by calculating the enormous economic value that would be washed down the drain with the excrement of horses and people. Twenty years later the editors of the Paris *Journal of Modern Chemistry* again took up a position against such a "public misdeed." During the middle of the last century, a sixth of the area of Paris produced fifty kilograms of fresh salads, fruits, and vegetables per capita, more than the 1980 level of per capita consumption. For each hectare of the Marais, 6.5 persons were employed full time gardening and scavenging, and more people were engaged in sales. During four decades, enough new "soil" was produced to expand the growing area by six percent per year. The growing techniques reached maximum sophistication in the 1880s: inter- and succession-cropping gave as many as six and never less than three harvests a year. Winter crops were made possible by the heat-fermentation of stable manure, bell-shaped glass cloches, special straw mats, and seven-foot-high walls surrounding the inner-city smallholdings. Kropotkin's claim, made in 1899, that Paris could supply London with green vegetables was by no means unreasonable. And since this system also produced more top soil than could be cultivated within the city if Paris humus was available for export, a proposal was made to have it collected on the streets by old-age

---

fully trustworthy. **Scott** gives information on the discovery of the sea as a place for bathing in the nineteenth century.

pensioners and to use the new railroads to enrich the countryside with this city product. Even after the first two modern aqueducts had been completed—one 81 kilometers long in 1865 and another 106 kilometers long in 1871—the use of water for the transport of excrement remained the rare exception.[41]

By the end of the third quarter of the century, two national ideologies concerning the value of sewers faced one another across the English channel. Victor Hugo gave literary expression to the French position. In his chapters on the city's innards in Les Misérables, the city of Paris is affected by incurable constipation. "No doubt," says he, "the sewer of Paris has been for millennia the city's disease, the open wound that festers at its bottom and that is part of a city's very nature": "l'égout et le vice que la ville a dans son sang." Any attempt to increase the foul matter that is stuffed down the drains could not but increase the already unimaginable horrors of the city's *cloaca*.

As testimony to the fantasy of cosmic constipation, these chapters are a work of art. But their references to the horrors of medieval sewers must not be read as historical information. In the Paris of the twelfth century each houseowner kept his drinking water in an open "tine" or vat, filled by a contracted water-carrier from two wooden pails that he carried by means of a "grouge"; since most wells gave brackish

---

41. For the physical setting of daily life in Paris, see **Farge 1979** and **1982**. Water was carried from fountains or peddled by sellers. During the mid-nineteenth century, public laundries in each neighborhood were modernized and covered with nonsupported roofs; there were 37 in 1848, 126 in 1860, and 422 by 1886. For the public controversy about sewers, see **Saddy, Jacquemet** and **Guillerme**. For sources and secondary literature on city gardening in Paris see **Stanhill**. For the history of water use in a region of France see **Thuillier 1968, 1969**. For the conflicting ideologies, **Beguin**. On the utopia of an odorless city, see **Gleichmann**. According to **Roche**, between 1700 and 1789 the total piped water reaching Paris doubled, but family consumption did not noticeably

water, drinking water was brought in from the Seine. During the thirteenth century when Paris was walled in, four public fountains were fed from an old aqueduct. There was no question of underground sewers. The "great sewer" to which a 1412 document refers was the brook of Menilmontant, which was not walled in until 1740 and not covered over until only a generation later. Its banks, just outside the city wall, were choice land for gardens. From the twelfth century on, the archbishop of Paris and the canons of Notre Dame grew their vegetables there. In the seventeenth century this brook fed the luscious Folie-Regnault, a preferred spot for the garden parties of the very rich. Both the physical reality and the literary topos of black waters crisscrossing an underground city are nineteenth-century creations. Clearly the imagery reflects the new anatomical vision of the body's innards.

The Polluting Sewers of London

On the other side of the channel the opposing view of the value of sewers was expressed in 1871 by the Prince of Wales, before he become King Edward VII. If he had not been the crown prince, he said, his next preference would be to become a plumber. To understand this enthusiastic endorsement of hydraulic engineering, it is necessary to remember that between 1848 and 1855 no fewer than six parliamentary

---

increase. The population increase and the monumental fountains in the Luxembourg Gardens and Les Tuileries swallowed it up. Roche estimates for 800,000 inhabitants of Paris in 1789: 300 tubs in public baths, 200 rental buckets and at best 1000 tubs in private houses. By the end of the century the fear of bad waters increases. **Muller** notes that from 1780 to 1789 bottled water became fashionable. At the end of the Empire fifty royal bureaus throughout France supervised their distribution. Their approval and the diagnosis of the right water for each client becomes a new expertise. In Paris twenty-two kinds of water were sold, in Lyons only seventeen. The cost per *pinte*—a large bottle—was equivalent to a laborer's daily wages.

commissions had been established to improve London sewage. But in spite of such efforts, conditions became intractable. The banks of the Thames between Waterloo and Westminster Bridge became covered with a thick accumulation of foul and offensive mud which was exposed at low tide. In 1849 and again in 1853-54, epidemics of Asiatic cholera took the lives of some 20,000 people. In the midst of the epidemic Parliament passed a new and more stringent act that rendered scavenging more effective to remove the night soil that was being generated by London masses. But the new pollution of the Thames was not primarily caused by them. It was due to the upper classes, who had installed the WC.

These devices had multiplied rapidly in London, partly because of the social status they conferred on the owners. By ordinance their contents had to end up in cesspools on the owner's own premises; but in spite of the veto, an increasing number of cesspools were connected to main sewers. These sewers at midcentury were mostly old watercourses, walled in but not covered as they flowed into town. Twenty years later British engineers were proudly on their way to improving the sanitary conditions of London without having to outlaw the WC. They had become the world leaders in the calculation, design, construction, maintenance, and ventilation of sewers for a population whose per capita consumption of water had reached levels that Paris would equal only two generations later. The future Edward VII simply voiced his admiration for the advanced technology of his time.

Waterworks Flush the U.S. Household

In the United States, the history of nineteenth-century waterworks has probably been less influenced by traditional attitudes toward water.[42] Most cities were settled by people

42. Besides the introduction to post-medieval water supply technologies in **Kennard**, **Rawlinson**, and **Daumas**, consult **Blake** on the social history of water on the east coast, for the United States. **Tarr** is a

from many European countries, each bringing their own attitude toward bathhouses, sociability in latrines, and cleanliness. Until the beginning of the nineteenth century all U.S. cities obtained their water from local sources: wells, cisterns, springs, and rivers; one to three gallons of water per day per person were mainly used for drinking, cooking, and laundry. Wide open patterns of settlement encouraged the establishments that used water for productive purposes to move outside town. Unlike their contemporaries in Europe, the majority of U.S. cities were built out of wood. The large fires at the beginning of the century led to demands for water to be used in fire fighting. By 1860 some 140 waterworks had been constructed. Technical breakthroughs facilitated these projects. From the time of the Romans, aqueducts had to be elevated when they passed over valleys. The U.S. cities that built waterworks during the last quarter of the nineteenth century no longer labored under this constraint. New, strong iron pipes had become available; these could withstand high pressures and therefore follow the profile of the ground. American cities that built waterworks for the prime purpose of firefighting were, from the beginning, concerned with water pressure, and the combination of the new iron pipes with available water pressure made it logical to deliver water right into homes. Wherever tap water reached the households, water consumption increased by a factor of between twenty and sixty, which meant that a rate of thirty to 100 gallons a day became typical.[43] The well-flushed home acquired a dominant position among American culture symbols.

---

valuable introduction to the social history of U.S. sewage. On the impact of water in housholds, see **Strasser** and **Van der Ryn**. On the origins of the ideology under which antibacterial defenses became U.S. policy, see **Temkin**.

43. During the nineteenth century, status is progressively tied to cleanliness. In many societies throughout history, some outsiders have been shunned as untouchables because of the impurity which they might com-

Piped flow per capita for the first time reached again Roman levels, but the distribution, in the United States, was incomparably more democratic. This new bounty was used in large part for the transport of waste. By the time of the Civil War, towns of more than 100,000 people had waterworks. By the end of the century, about 3000 waterworks were in operation, serving most of the towns with a population of 2500 or more. But this does not mean that water immediately reached many households. We have data for Muncie, Indiana. In 1890 between one-eighth and one-fifth of Muncie families had at least the crudest access to running water, perhaps a hydrant in the yard or a faucet in the iron sink outside the door. Most pumped their water from wells in the backyard. At the beginning of the Civil War, toilets or bathtubs were considerable luxuries. The water department charged extra for houses sporting them. In 1893 four-fifths of all inhabitants of Baltimore had access to an outdoor privy only; in New York, almost half had indoor privies. In 1866 only one-eighth of Chicago was served by sewers. Water and slops were mostly carried by women. Only after World War I did bathrooms cease to be luxuries. In the four years from

---

municate; during the nineteenth century the lower classes came to be seen not as *impure* but as *unclean*. Balzac's novels are a rich source for documenting the transition; see **Pfeiffer**. The "strong and wild odor of the peasants" becomes unsupportable to the refined nose of a lady. The refined lady radiates "a charming perfume of the bourgeoisie." The scrubbed and educated classes are aware of the smell of corruption on the lower levels; see **Perrot**. The harlot is perceived by analogy with the sewer that drains the decent home of disturbing forms of male lust. To rise in society means to become clean and to live in a decently cleaned house. But in Europe, especially on the continent, the accent lay primarily on the dusting and scrubbing of the home inside. For Switzerland, see **Heller**. In America the whole house, one's own and that of others, has become a uniquely dominant culture symbol. **Cohn** examines what people say the house is and what they admire about the house and speaks of the house as a mirror in which they observe the values of their own

1921 to 1924, they doubled throughout the whole of the United States. A national survey in the late 1920s showed that seventy-one percent of urban and thirty-three percent of rural families had installed bathrooms in their homes. Water now mostly served washing, cleaning, and flushing.[44]

At first this waste water was directed into cesspools and privy vaults. Around 1880 this gave rise to an unanticipated event throughout U.S. cities. Everywhere the capacity of cesspools was overwhelmed; the surrounding soil no longer could absorb the water. The government of Rhode Island identified as the major health problem the fact that residents had introduced more water into their dwellings than they could get rid of. Benjamin Lee, secretary of health in Pennsylvania, warned that "copious water supplies constitute a means of distributing foul pollution over immense areas and constitute a nuisance prejudicial to public health."

The WC Integrates U.S. Culture

The fundamental economics of water circulation became visible.[45] Under both public and private management, the

---

culture. Around the turn of the century in the United States, cleanliness put all other qualifications for a desirable house into the shade. Through the acquisition of cleanliness, minorities could merge and dissolve into the mainstream culture. For this see also **G. Wright**. Musings on the significance of the great cleanup of the nineteenth century are in **Enzensberger.**

44. Information from **Strasser**, *passim*. The change is reflected in language: **Mencken**, *The American Language*, Supplement I, 639–41. About 1870 the latrine disappears. Toilet, retiring room, washroom, comfort station appear. *Pissoir* is still indecent in the United States, rare in England, and called *vespasienne* in France. Powder room (originated by some learned speakeasy proprietor to designate the ladies' retiring room.) Other words: restroom, dressing room, ladies room, cloakroom, lavatory. *OED* traces "toilet" to 1819, "in US esp."

45. "The city of Leonia refashions itself every day.... On the sidewalks, encased in spotless plastic bags, the remains of yesterday's

cost of getting rid of abundant water proved many times more costly than getting it there in the first place. This disproportion was increased further when many large U.S. cities decided to combine the sewers for waste with storm sewers for rain. This decision implied the construction of sewers with a capacity to carry out of the city a much larger amount of water than was carried in and to provide a margin for error so that sewage would not float through streets not designed for this purpose when an occasional heavy rain overwhelmed the available capacity.

A second unpremeditated factor increased the cost of sewers. Engineers relied on the dilution and dispersal of waste in natural bodies of water as though this were the same as their disposing of them completely. They realized only slowly that the same toilet-dependent ideology of cleanliness that produced artificial marshes around cesspools near dwellings also polluted rivers and delivered souvenirs from the cities upstream to the waterworks downstream. By the end of the century, the spread of fecal-borne infection through tap water became common. The circulation of water became a major agent in the circulation of disease. Engineers were faced with the choice of applying their always limited resources either to the treatment of sewage before its disposal or to the treatment of water supplies. For the first half of this century, they chose to sterilize the water supplies, even though filtering systems and chemical treatment, mainly with chlorine, became increasingly expensive.

One reason they chose "water purification" was probably the hold that Dr. Koch's discoveries had gained over the fan-

---

Leonia await the garbage truck. . . . Nobody wonders where, each day, they carry their load of refuse. Outside the city, surely; but each year the city expands. . . . A fortress of indestructible leftovers surrounds Leonia, dominating it on every side, like a chain of mountains. This is the result: the more Leonia expels goods, the more it accumulates them; the scales of

tasies of voters at the end of the last century. His theories of bacteriology tended to replace the old filth-theory of corrupting emanations with a new germ-theory that seemed to explain the appearance of specific diseases. Instead of contact with foul airs, bodily invasion by microbes was the thing to be avoided. Citizens demanded, above all, to be supplied with "germless drinking water" when they opened their taps. During the first half of the twentieth century, several generations of Americans learned to abstain from drinking water unless it came from an approved faucet or bottle. Bathing in unchlorinated brooks and drinking from untested fountains became, for many people, a memory of scouting during childhood or the reminder of a romantic past.

Recovery of "Stuff"

Then, during the second half of the twentieth century, what came from the faucet ceased to be odorless. Its content of entirely new and unthought-of pollutants became known. Many people refused to serve it to their children as a drink. The transformation of H_2O into a cleaning fluid was complete. In the imagination of the twentieth century, water lost both its power to communicate by touch its deep-seated purity and its mystical power to wash off spiritual blemish. It has become an industrial and technical detergent, feared both as a poisonous stuff and as a corrosive for the skin. During the last years of the Carter presidency, the cost of sewage treatment and collection had become the greatest expense that local governments foresaw during the 1980s. Only education costs the taxpayer more.

Water throughout history has been perceived as the stuff

---

its past are soldered into a cuirass that cannot be removed. As the city is renewed each day, it preserves all of itself in its only definitive form...." (**Calvino**, 114–15).

which radiates purity: H_2O is the new stuff, on whose purification human survival now depends. H_2O and water have become opposites: H_2O is a social creation of modern times, a resource that is scarce and that calls for technical management. It is an observed fluid that has lost the ability to mirror the water of dreams. The city child has no opportunities to come in touch with living water. Water can no more be observed; it can only be imagined, by reflecting on an occasional drop or a humble puddle.

I believe these reflections to be relevant to the decision on Town Lake that has to be made in Dallas.

Bibliography

Alexander, Samuel. 1966. *Space, time and deity.* 2 vols. New York: Dover Publications.

Aptorowitzer, O. 1928. Die Paradiesesflüsse des Kurans. *Monatsschrift für Geschichte und Wissenschaft des Judentums* 72, new ser. 36:151–55.

Ardener, Shirley, ed. 1981. *Women and space: Ground rules and social maps.* London: St. Martin's Press.

Ariès, Philippe. 1982. *The hour of our death.* Trans. Helen Weaver. New York: Random House. Originally published as *L'homme devant la mort.* Paris: Seuil, 1977.

Aron, Jean Paul. 1967. Essai sur la sensibilité alimentaire à Paris au XIXe siècle. *Cahiers des Annales* 25. Paris: Armand Colin.

Bachelard, Gaston. 1969. *The poetics of space.* Trans. Maria Jolas. Boston: Beacon Press. Originally published as *La Poétique de l'espace.* Paris: Presses universitaires de France, 1957.

———. 1983. *Water and dreams: An essay on the imagination of matter.* Trans. E. R. Farrell. Dallas: The Dallas Institute of Humanities and Culture. Originally published as *L'Eau et les Rêves.* Paris: Corti, 1942.

Bächtold-Stäubli, H. 1936. *Handwörterbuch des deutschen Aberglaubens.* Berlin: DeGryter.

Barrett, C. K. 1962. *The gospel according to St. John.* London: SPCK.

Beguin, F. 1977. Les Machines anglaises du confort. *La Recherche* 29:155–86.

Behm, Johannes. 1938. Koilía. *Theologisches Wörterbuch zum Neuen Testament.* Ed. Gerhard Kittle. Vol. 3. Stuttgart, 786–89.

Belgrand, E. 1872–1875. *Les aqueducts romains.* 5 vols. Paris.

Benveniste, Emil. 1969. *Le Vocabulaire des institutions indo-européennes.* 2 vols. Paris: Editions de Minuit.

Bertier, A. G. 1921–1922. Le mécanisme cartésien et la physiologie au XVIIe siécle. *Isis* 7:21–58.

Beylebyl, J. J. 1974. The growth of Harvey's "de Motu Cordis." *Bulletin of the History of Medicine* 47:427–70.

Bidez, Joseph, and F. Cumont. 1938. *Les Mages hellénisés.* 2 vols. New York.

Ivan Illich

Blake, Nelson M. 1956. *Water for the cities: A history of the urban supply problem in the USA*. Syracuse: Syracuse University Press.

Blaser, Robert. 1960. *Theophrastus von Hohenheim, genannt Paracelsus. Liber de Nymphis, Sylphis, Pygmaeis et Salamandris et de ceteris spiritibus*. Bern: Franke.

Boccaccio, Giovanni. 1951. *Genealogiae deorum gentilium libri*. Bari: Ricciardi.

Boisemard, M. E. 1958. De son ventre couleront des fleuves d'eau. *Revue Biblique* 65:523-46.

——— 1959. De son ventre couleront des fleuves d'eau: Les citations targumiques dans le quatrième évangile. *Revue Biblique* 66:374-78.

Bollnow, Otto. 1971. *Mensch und Raum*. 2nd ed. Stuttgart: Kohlhammer.

Bonaparte, Marie. 1946. The legend of the unfathomable waters. *American Imago*. 20-31.

Bonnet, H. 1925. Die Symbolik der Reinigung im aegyptischen Kult. *Angelos* 1:103-121.

Borie, Jean. 1973. *Le tyran timide: Le naturalisme de la femme au XIXe siecle*. Paris: Klincksieck.

Borzsak, I. 1951-1952. Aquis submersus. *Acta Antiquae Academiae Scientiarum Hungaricae, Budapest* 1:200-24.

Bossel, Hartmut, ed. 1982. *Wasser: wie ein Element verschmutzt und verschwendet wird*. Frankfurt: Fischer Alternative.

Boughali, Mohammed. 1974. *La Représentation de l'espace chez les marocains illettrés: Mythes et tradition orale*. Paris: Anthropos.

Braun, F. M. 1949. L'eau et l'Esprit. *Revue Thomiste* 49:5-30.

Brown, Raymond E. 1966. *The gospel according to St. John*. New York: Anchor Books.

Brunot, Ferdinand. 1966. *Histoire de la langue française des origines à nos jours*. Vol. 6, pt. 1. Paris: Armand Colin, 176ff.

Buck, C. D. 1949. *A dictionary of selected synonyms in the principal Indo-European languages*. Chicago: University of Chicago Press.

Budge, Wallis. [1909] 1980. *The liturgy of funerary offerings: The Egyptian texts with English translation*. Reprint. New York: Arno Press.

Buffet, B., and R. Evrard. 1950. *L'eau potable à travers les âges*. Liége.

Cabrol, F. 1926. Eau: usage de l'eau dans la liturgie. Eau bénite. *DACL* 4:1680-690.

Bibliography

Caillois, R. 1938. *Le mythe et l'homme*. Paris: Gallimard.

Calvino, Italo. 1975. *Invisible cities*. Trans. William Weaver. London: Secker and Warburg. Originally published as *Le Città invisibili*. Rome: Einaudi, 1972.

Cayrol, Jean. 1968. *De l'espace humain*. Paris: Seuil.

Chadwick, Edwin. 1887. *The health of nations*. 2 vols. Ed. R. W. Richardson. London.

Chaunu, Pierre. 1974. *Histoire science sociale: La durée, l'espace et l'homme à l'époque moderne*. Paris: Société d'Edition de l'Enseignement Supérieur. See chapter 2.

Choay, Francis. 1974. La ville et le domain bâti comme corps dans les textes des architects théoriciens de la première renaissance italienne. *Nouvelle Revue de Psychanalyse* 9:229-54.

Cohn, Jan. 1979. *The palace and the poorhouse: The American house as a cultural symbol*. Lansing: Michigan State University.

Clark, Kenneth. [1950] 1970. *The nude: A study in ideal form*. Reprint. London: Penguin.

———. 1980. *Feminine beauty*. London: Weidenfeld and Nicolson.

Corbin, Alain. 1982. *Le miasme et la jonquille: L'odorat et l'imaginaire social XVIIIe-XIXe siècles*. Paris: Aubier.

Corbin, Henri. 1977. Imago templi face aux normes prophanes. In *Eranos Jahrbüch—1974*. Leiden: E J. Brill.

Dange, Sadashiva Ambadas. 1979. *Sexual symbolism from Vedic ritual*. New Delhi: Ajanta Publications.

Danielou, J. 1958. Le symbolisme de l'eau vive. *Revue des Sciences religieuses* 32:335-46.

Daumas, Maurice, ed. 1968. *Histoire générale des techniques*. Vol. 3. Paris: Presses universitaires de France.

Degas, Edgar. 1970. *L'Opera completa*. Ed. Franco Russoli and Fiorella Minervino. Rome: Rizzoli.

Delcourt, Marie. 1964. *Pyross et Pyrrha: Recherches sur les valeurs du feu dans les légendes helléniques*. Paris: Belles Lettres.

Deonna, W. 1939. Croyances antiques et modernes: L'odeur suave des dieux et des élus. *Genava* 17:167-262. Exhaustive collection of sources and references.

Desaivre, Leo. 1898. Le mythe de la mère Lusine: Etude critique et bibliographique. *Mémoires de la Société de Statistique, Sciences, Lettres et Arts du département des Deux-Sèvres* 20.

Detienne, Marcel. 1972. *Les Jardins d'Adonis: La mythologie des aromates en Grèce.* Paris: Gallimard.

Deubner, L. 1942. Oedipusprobleme. *Abhandlungen der Akademie der Wissenschaften zu Berlin* 4.

Dickinson, Henry W. 1959. *The water supply of greater London.* London: Newcome Society.

Dockes, Pierre. 1969. *L'espace dans la pensée économique du XVIe au XVIIIe siècle.* Paris: Flammarion.

Dodd, C. H. 1954. *The interpretation of the fourth gospel.* Cambridge: Cambridge University Press.

Dodds, E. R. 1951. *The Greeks and the irrational.* Berkeley: University of California Press.

Duden, Barbara. 1984. Fluss und Verstockung: Praxis Johannes Storch. Doctoral dissertation, Technische Universität, West Berlin.

Durkheim, E. 1976. *Elementary forms of the religious life.* London: Allen & Unwin. Originally published as *Les formes élémentaires de la vie religieuse.* Paris: Presses universitaires de France, 1960.

Eliade, Mircea. 1964. *Traité de l'histoire des religions.* Paris: Payot.

Elias, Norbert. 1960. *Uber den Prozess der Zivilisation.* 2 vols. Munich: Francke.

Eliot, T. S. 1963. *The family reunion.* London: Faber.

Ernout, Alfred, and Antoine Meillet. 1967. *Dictionnarie étymologique de la langue latine: histoire des mots.* Paris: Klincksieck.

Enzensberger, Christian. 1968. *Smut.* London: Marion Boyars.

Fabricant, C. 1979. Binding and dressing nature's loose tresses: The ideology of Augustan landscape design. *Studies in eighteenth-century culture* 8.

Farge, Arlette. 1979. *Vivre dans les rues à Paris au XIXe siècle.* Paris: Gallimard.

———. 1983. L'espace parisien au XVIIIe siècle. *Ethnologie Française* 3.

Bibliography

Farwell, Béatrice. 1972. Courbet's "Baigneuses" and the rhetorical feminine image. *Art News Annual* 38:65-79.

Faure, A. 1978. Class malpropre, classe dangereuse? Quelques remarques à propos des chiffonniers parisiens au XIXe siècle. *L'haleine des faubourgs. Ville, habitat et santé au XIXe siècle.* Fontenay-sous-Bois: Recherche.

Filliozat, Jean. 1969. Le temps et l'espace dans les conceptions du monde indien. *Revue de Synthèse* 90.

———. 1974. *Laghu Prabandhan: Choix d'articles d'indologie.* Reprint. Leiden: E. J. Brill, 212-32.

Finer, S. E. 1967. *The life and times of Sir Edwin Chadwick.* London: Methuen.

Flourens, P. 1954. *Histoire de la découverte de la circulation du sang.* Paris.

Foizil. M. 1974. Les attitudes devant la mort au XVIIIe siècle: sépultures et suppressions des sépultures dans le cimetière parisien des Saints Innocents. *Revue Historique* 51:303-330.

Forbes, R. J. 1964. Water supply. *Studies in Ancient Technology* 1:149-94. Leiden: E. J. Brill.

———. 1965. Cosmetic and perfumes in antiquity. *Studies in Ancient Technology* 3:1-50. Leiden: E. J. Brill.

Franz, Marie-Louise. 1983. *Shadow and evil in fairy tales.* Dallas: Spring Publications.

Frazer, J. G. 1933. *The fear of the dead in primitive religion.* 3 vols. London: Macmillan.

Freud, Sigmund. 1961. Civilization and its discontents. *The Standard Edition of the Complete Psychological Works* Vol. 21. Trans. and ed. James Strachey. London: The Hogarth Press and The Institute of Psychoanalysis, 59-145.

Foucault, Michel. 1976. *The order of things.* London: Tavistock.

Frisch, O. R. 1957. Parity is not conserved, a new twist in physics? *Universities Quarterly* 2:235-44.

Frobenius, Leo. 1933. *Kulturgeschichte Afrikas.* Zürich: Phaidon.

Fustel de Coulanges. [1869] 1972. *La cité antique.* Reprint. Paris: Hachette.

Gaiffe, F. A. 1924. *L'envers du Grand Siècle: Etude historique et anecdotique.* Paris: Albin Michel.

Gaillard, Jean. 1958. s.v. Eau. *Dictionnaire de Spiritualité* Vol. 5. Paris: Beauchesne

Gay, Harold Fansworth. 1940. Sewage in ancient and medieval times. *Sewage Work Journal* 12:939-48.

Giedion, Siegfried. 1948. *Mechanization takes command: A contribution to anonymous history*. New York: Norton.

Ginzberg, Louis. 1968. *The legends of the Jews*. Philadelphia: Jewish Publication Society of America.

Glacken, Clarence J. 1967. *Traces on the Rhodian shore: Nature and culture in western thought from ancient times to the end of the eighteenth century*. Berkeley: University of California Press.

Gleichmann, P. R. 1982. Des villes propres et sans odeurs. *Urbi*. 88-100.

Goppelt, Leonard. 1969. s.v. hydor. *Theologisches Wörterbuch zum Neues Testament* 8:313-33.

Graham, V. E. 1959. Water imagery and symbolism in Proust. *Romanic Review* 50:118-28.

Grandquist, Hilma. 1965. Muslim death and burial. *Commentationes Humaniorum Litterarum*. 1-128.

Greverus, Ina Maria. 1972. *Der territoriale Mensch. Ein literaturanthropologischer Versuch zum Heimatphaenomen*. Frankfurt am Mainz: Athenaeum.

Guillerme, A. n.d. Quelques problèmes d'eau dans les villes du bassin parisien au Moyen Age. 2 vols. Thesis, Ecole des Hautes Etudes de Sciences Sociales, Paris.

Guillerme, Jacques. 1977. Le malsain et l'économie de la nature. *XVIIIe Siècle 9*.

Griaule, Marcel. 1966. *Dieu d'Eau: Entretiens avec Ogotemeli*. Paris: Fayard. Translated into English as *Conversations with Ogotemeli: An introduction to Dogon religious ideas*. Oxford: Oxford University Press, 1975.

Hahn, Ingrid. 1963. Raum und Landschaft in Gottfrieds Tristan. Ein Beitrag zur Werkdeutung. *Medium Aevum 3*.

Haight, Elizabeth Hazelton. 1958. *The symbolism of the house-door in classical poetry*. New York: Longmans Green.

Heller, G. 1979. *"Propre en ordre." Habitation et vie domestique 1850-1930. L'exemple Vaudois*. Lausanne: Editions en Bas.

Hopkins, E. Washburn. 1975. *Epic Mythology.* 2d. ed. Benares: Motilal.

Hupping Stoners, Carol, ed. 1977. *Goodbye to the flush toilet.* Emmaus, Pennsylvania: Rodale.

Husserl, Edmund. 1969. *Formal and Transcendental Logic.* Trans. Dorion Cairns. The Hague: Nijhoff.

Hvidberg, Flemming Friis. 1962. *Weeping and laughter in the Old Testament.* Leiden: E. J. Brill.

Ibn al-Nafiz. 1968. *The Theologus Autodidactus.* Ed. J. Schacht. Oxford: Oxford University Press.

Illich, Ivan. 1982. *Gender.* London: Marion Boyars.

———. 1984. *Schule ins Museum: Phaidros und die Folgen.* Bad Heilbrunn: Klinkhardt.

Jacquemet, Gérard. 1979. Urbanisme parisien: La bataille du tout-à-l'égout à la fin du XIXe siècle. *Revue d'Histoire moderne et contemporaine.* 505-48.

Jordanova, C. J., and R. S. Porter. 1979. *Images of the earth: Essays in the history of environmental science.* British Society of Environmental Science Monographs 1. London: Giles.

Jammer, Max. 1954. *Concepts of space.* Cambridge, Mass.: Harvard University Press.

Jolas, Tina. 1977. Parcours cérémonial d'un terroir villageois (Minot). *Ethnologie française* 7:7-28.

Jüthner, S. 1950. s. v. Bad. *Reallexikon für Antike und Christentum* 30.

Karnoch, C. 1972. L'étranger ou le faux inconnu: Essai sur la définition spatiale d'autrui dans un village lorrain. *Ethnologie Française* 1-2:107-22.

Kennard, J. 1958. Sanitary engineering: Water supply. *A history of technology.* Vol. 4. Ed. C. Singer. Oxford: Oxford University Press, 489-503.

Kerényi, Karl. 1942. Hermes der Seelenführer. Das Mythologem vom männlichen Lebensursprung. In *Eranos Jahrbuch* 9. Zürich: Rhein Verlag, 9-63.

———. 1976. *Die Mythologie der Griechen.* 2 vols. Munich: DTV.

Kittel, Gerhard, ed. 1933-1973. *Theoligisches Wörterbuch zum Neuen Testament.* 9 vols. Stuttgart.

Kramrich, Stella. 1963. The triple structure of creation in the Rig Veda. *History of Religions*. Vol. 2, 140-75, 257-85.

———. 1968. *Unknown India: Ritual art in tribe and village*. Philadelphia: Philadelphia Museum of Art.

Kohler, J. 1895. *Der Ursprung der Melusinensage*. Leipzig.

Krauss, Rosalind. 1967. Manet's nymph surprised. *Burlington Magazine* 109:622-27.

Kugelmann, Robert. 1982. *The windows of soul: Psychological physiology of the human eye and primary glaucoma*. London: Associated University Presses.

Ladendorf, H. 1957. Der Duft in der Kunstgeschichte. In *Festschrift E. Meyer*. Hamburg: Hauswedell, 251-73.

Lagrange, M. J. 1903. *Etude sur les religions sémitiques*. Paris.

Landolt, Hermann. 1977. Sacralraum und mystischer Raum in Islam. In *Eranos Jahrbüch 44—1975*. Leiden: E. J. Brill..

Laporte, Dominique. 1979. *Histoire de la merde*. Paris.

Larousse, Pierre. 1876. s.v. parfum. *Grand dictionnaire universel du XIXe siècle* 12.

Latte, Kurt. 1968. *Kleine Schriften zu Religion, Recht, Literature und Sprache der Griechen und Römer*. Munich: Beck.

Lawrence, Roderick J. 1982. Domestic space and society: A cross cultural study. *Comparative Studies of Society and History*. 104-30.

Lebeuf, J. P. 1961. *L'habitation des Fali*. Paris: Hachette.

Leclercq, H. 1924. s. v. bénitier. *Dictionnaire d'Archéologie Chrétienne et de Liturgie* Vol. 2, 758-71.

Lefevre, A. 1950. La blessure du côté. *Etudes Carmélitaines*. Paris: Le Coeur, 109-22

Lepenies, Wolf. 1963. *Melancholie und Gesellschaft*. Freiburg: Suhrkamp.

Lewy, H. 1929. Sobria Ebrietas. Untersuchungen zur Geschichte der antiken Mystik. *Beihefte zur Zeitschrift für neutestamentliche Wissenschaft*.

Liebert, Gösta. 1976. *Iconographic dictionary of the Indian religions*. Leiden: E. J. Brill.

Lilja, Saara. 1972. *The treatment of odours in the poetry of antiquity*. Helsinki.

Lincoln, Bruce. 1982. Waters of remembrance and waters of forgetfulness. *Fabula* 23:19–34.

Lippe, Rudolf zur. 1981. *Naturbeherrschung am Menschen*. Vol. I. Geometrisierung des Menschen und Repräsentation des Privaten im französischen Absolutismus. Frankfurt: Syndikat.

———. 1982. *Die Geometrisierung des Menschen*. Ausstellungskatalog. Oldenburg: Bibliothekes und Informationssystem der Universitkat Oldenburg.

Lohmeyer, Ernst. 1919. Vom göttlichen Wohlgeruch. *Heidelberger Akademie der Wissenschaften*. Philos. Hist. Klasse 10.

Ludwig, A. 1920. Schlemihl. *Archiv für das Studium der neueren Sprachen und Literaturen*. Deutsches Sonderheft.

———. 1921. Nachträge zu Schlemihlen.

Lüers, Grete. 1926. *Die Sprache der deutschen Mystik des Mittelalters im Werke der Mechthild von Magdeburg*. Munich: Reinhardt.

Luther, W. 1935. *"Wahrheit" und "Lüge" im ältesten Griechentum*. Leipzig: Boana.

McGovern, J. J. 1959. The waters of death. *Biblical Quarterly* 21:350–58.

Mahnke, Dietrich. 1937. *Unendliche Sphäre und Allmittelpunkt*. Halle.

Malraux, André. 1926. *La tentation de l'Occident*. Paris: Gallimard.

Mandrou, Robert. 1974. *Introduction à la France moderne 1500–1640. Essai de psychologie historique*. 2d. ed. Paris: Albin Michel.

Mani, Vettam. 1975. *Puranic Encyclopedia*. Benares: Motilal.

Maringer, Johannes. 1973. Das Wasser im Kult und Glauben der vorgeschichtlichen Menschen. *Anthropos* 68:705–76.

Marmier, Claire Kersaint. 1947. *La mystique des eaux sacrées dans l'antique Armor: Essai sur la conscience mythique*. Paris: Vrin.

Martin, Roland. 1956. *L'urbanisme dans la Grèce antique*. Paris: Picard.

Maurmann, Barbara. 1976. *Die Himmelsrichtungen im Weltbild des Mittelalters*. Munich: Fink.

Meister, Kurt. 1925. *Die Hausschwelle in Sprache und Religion der Römer*. Heidelberg: Winter.

Menard, Jacques E. 1955. L'interprétation patristique de Jean 7, 38. *Revue de l'Université de Ottawa*. Sec. Spéciale 25:5-25.

Mencken, H. L. 1963. *The American language*. London: Routledge & Kegan Paul.

Meyer, Rudolf, and Friedrich Hauck. 1938. s. v. katharos. *Theologisches Wörterbuch zum Neuen Testament*. Ed. Gerhard Kittel. Vol. 3. Stuttgart, 416–33.

Meyerhof, M. 1933. Ibn al-Nafis und seine Theorie des Lungenkreislaufes. *Quellen und Studien zur Geschichte der Naturwissenschaften und der Medizin* 4:37–88.

Monier-Williams, M. [1899] 1963. *A Sanscrit-English dictionary etymologically and philologically arranged*. Reprint. Oxford: Clarendon Press.

Mounin, Georges. 1965. Essai sur la structuration du lexique de l'habitation. *Cahiers de Lexicologie* 6:9–24.

Muller, P. 1975. Les eaux minérales en France à la fin du XVIIIe siècle. Mémoire de maîtrise, University of Paris.

Mumford, Lewis. 1961. *The city in history: Its origins, its transformation and its prospects*. London: Secker and Warburg.

Murard, Lion, and Patrick Zylberman, eds. 1978. *L'haleine des faubourgs: Ville, habitat et santé au XIXe siècle*. Fontenay-sous-Bois: Recherche.

Murray, J. A. H. 1973. *The shorter OED on historical principles*. 3d ed. Oxford: Clarendon Press.

Nestle, E. 1906. Der süsse Geruch als Erweis des Geistes. *Zeitschrift für Neutestamentliche Wissenschaft und Kunde des Urchristentums* 7:95–96.

Negrier, Paul. 1925. *Les bains à travers les âges*. Paris.

Niangoran Bouah, G. 1964. *La division du temps et le calendrier rituel des peuples lagunaires de Côte d'Ivoire*. Travaux et Mémoires de l'Institut d'Ethnologie 68. Paris.

Nicolas, G. 1966. Essai sur les structures fondamentales de l'espace dans la cosmologie Hausa. *Journal de la Société Africaniste* 36.

Nicolson, Marjorie Hope. 1950. *The breaking of the circle*. Evanston: Northwestern University Press.

———. 1966. The discovery of space. *Medieval and Renaissance Studies* 1:40–59.

Niessen, H. 1868. *Templum*. Berlin.

Ninck, Martin. [1921] 1967. *Die Bedeutung des Wassers im Kult und Leben der Alten. Eine symbolgeschichtliche Untersuchung*. Reprint. Wiesbaden: Wissenschaftliche Buchgesellschaft.

Nohain, Jean, and F. Caradec. 1976. *Le pétomane*. Los Angeles: Sherbourne Press.

Norden, Eduard. 1939. *Aus altrömischen Priesterbüchern*. Leipzig: Harassowitz.

Notopoulos, James A. 1938. Mnemosyne in oral literature. *Transactions and Proceedings of the American Philological Association* 69:465-93.

Novotny, Karl A. 1969. *Beiträge zur Geschichte des Weltbildes: Farben und Weltrichtungen. Wiener Beiträge zur Kulturgeschichte und Linguistik* 17. Vienna: F. Berger.

Ohly, Friedrich. 1977. Geistige Süsse bei Ottfried. *Schriften zur mittelalterlichen Bedeutungsforschung*. Darmstadt: Wissenschaftliche Buchgesellschaft.

Olivetti, Marco. 1967. *Il tempio simbolico cosmico*. Rome: Edizioni Abete.

Ong, Walter J. 1982. *Orality and literacy: The technologizing of the word*. London: Methuen.

Ott, Sandra. 1980. Blessed bread, "first neighbours" and assymetric exchange in the Basque country. *Archives Européennes de Sociologie*.

Pagel, W. 1944. The religious and philosophical background of van Helmont's science and medicine. *Supplement to the Bulletin of the History of Medicine* 2.

Palmer, R. 1973. *The water closet: A new history*. Newton Abbott: David & Charles.

Parker, Robert. 1983. *Miasma. Pollution and purification in early Greek religion*. Oxford: Clarendon Press.

Patte, Pierre. 1769. Sur la distribution vicieuse des villes. *Mémoire sur les objets les plus importants de l'architecture*. Paris. 28f.

Pauly, August F. von. 1964. *Der kleine Pauly. Lexikon der Antike*. 5 vols. Ed. Konrad Ziegler. Stuttgart: A. Druckmüller Verlag.

Peabody, Berkley. 1975. *The winged word: A study in the technique of ancient Greek oral composition as seen principally through Hesiod's "Works and Days."* Albany: State University of New York Press.

Perrot, Philippe. 1981. *Les dessus et les dessous de la bourgeoisie*. Paris: Fayard.

———. 1984. *Le travail des apparences: ou les transformations du corps féminin. XVIIIe et XIXe siècles*. Paris: Seuil.

Ivan Illich

Peters, Hans Albert. 1971–1972. *Halbe Unschuld. Weiblichkeit um 1900. Europäische Graphik aus der Zeit des Jugendstils.* Cologne: Wallraf-Fischart Museum.

Peuckert, W. E. 1953. Traufe und Flurgrenze. *Zeitschrift für Volkskunde.* 66–83.

Pfeiffer, Charles Leonard. 1949. *Taste and smell in Balzac's novels.* Phoenix: University of Arizona.

Pingaud, Marie Claude. 1973. Le langage de l'assolement (Minot). *Homme* 13:163–75.

Pokorny, Julius. 1959. *Indogermanisches etymologische Wörterbuch.* Bern: Francke.

Poulet, Georges. 1961. *Les métamorphoses du cercle.* Paris: Plon.

Prado, G. G. 1969. Review of analytical philosophy of knowledge. *Dialogue* 8:503–07.

Quiguer, Claude. 1979. *Femmes et machines de 1900: Lecture d'une obsession Modern Style.* Paris: Klincksieck.

Rahner, Hugo. 1941. Flumina de Ventre Christi: Die patristiasche Auslegung von Joh 7, 37–38. *Biblica* 22:269–301, 367–403.

———. 1963. *Greek myths and Christian mystery.* New York: Harper & Row. Originally published as *Griechische Mythen in christlicher Deutung.* Zürich: Rhein Verlag. 1957.

———. 1964. *Symbole der Kirche.* Salzburg: O. Meuller. 117–238.

Rawlinson, J. 1958. Sanitary engineering: Sanitation. In *A History of Technology.* Ed. C. Singer. Vol. 4. Oxford: Oxford University Press, 504–519.

Rhode, Erwin. 1898. *Psyche: Seelencult und Unsterblichkeitsglaube der Griechen.* 2d. ed. 2 vols.

Robins, R. W. 1946. *The story of water supply.* Oxford: Oxford University Press.

Robinson, J. A. T. 1962. The significance of foot washing. In *Festschrift O. Cullmann.* 144–47.

Roche, Daniel. 1984. Du temps de l'eau rare du moyen âge à l'époque moderne. *Annales ESC.* Vol. 19, pt. 2:383–399.

Rogers, Susan C. 1979. Espace masculin, espace féminin: Essai sur la différence. *Etudes rurales* 74:87–110.

Bibliography

Roumeguere-Eberhardt, J. 1957. La notion de vie: Base de la structure sociale Venda. *Journal de la société des Africanistes* 27, fasc. 11. Paris.

———. 1963. Pensée et société africaines. In *Cahiers de l'Homme*. Paris and The Hague: Mouton.

Ruberg, Uwe. 1965. *Raum und Zeit im Prosa-Lancelot*. Munich: Fink.

Russoli, Franco. 1970. *L'opera completa di Degas. Apparati critici e filologici di Fiorella Minervino*. Rome: Rizzoli Editore.

Rykwert, Joseph. 1972. *On Adam's house in paradise: The idea of the primitive hut in architectural history*. New York: The Museum of Modern Art.

———. 1976. *The idea of a town: The anthropology of urban form in Rome, Italy and the ancient world*. London: Faber and Faber.

Ryle, Gilbert. 1950. *The concept of mind*. New York: Barnes and Nobel.

Sabine, E. L. 1934. Latrines and cesspools in medieval London. *Speculum* 9:303–21.

———. 1937. City cleaning in medieval London. *Speculum* 12:19–43.

Saddy, Pierre. 1977. Le cycle des immondices. *XVIIIe Siècle*. 203–214.

Sardello, Robert J. 1983. The suffering body of the city: Cancer, heart-attack and herpes. *Spring: An Annual of Archetypal Psychology and Jungian Thought*. 145–164.

Schacht, J. 1957. Ibn al-Nafis, Servetus and Colombo. *Al-Andalus* 22:317–36.

Schneider, U. 1961. Die alt-indische Lehre vom Kreislauf des Wassers. *Saeculum* 12:1–11.

Schmidt, N. 1966. Niemandsland. In *Volksglaube und Volksbrauch: Gestalten, Gebilde, Gebärden*. Berlin: Schmidt Verlag.

Schulz, Hans, and Otto Basler. 1983. *Deutsches Fremdwörterbuch*. Vol. 6. Berlin: Walter Gruyter.

Scott, George Ryley. 1939. *Story of baths and bathing*. London.

Segel. Harold B. 1974. *The baroque poem*. New York: Dutton.

Sertillanges, L. D. 1945. *L'idée de création et ses retentissements en philosophie*. Paris: Desclé.

Sextus Julius Frontinus. 1913. *The two books on the water supply of the city of Rome*. Trans. Clemens Herschel. London: Longmans Green.

Simmel, Georg. 1953. Soziologie des Raumes. *Schmollers Jahrbuch* 27.

———. 1959. Philosophie der Landschaft. *Brücke und Tür: Essays des Philosophen zur Geschichte, Religion, Kunst und Gesellschaft.* In collaboration with Margarete Susman. Stuttgart: Landmann.

Stafford, Barbara Maria. 1939. *Voyage into substance: Art, science, nature and the illustrated travel account, 1760-1840.* Cambridge: The MIT Press.

Stanhill, G. 1977. An urban agro-ecosystem: The example of nineteenth-century Paris. *Ecosystems* 3:269-84.

Staudacher, W. 1942. *Die Trennung von Himmel und Erde: Ein vorgriechischer Schöpfungsmythos bei Hesiod und den Orphikern.* Thübingen.

Stith-Thompson. 1966. *Motif index of folk tale literature.* Bloomington: Indiana University Press.

Strack H., and P. Billerbeck. 1926. *Kommentar zum neuen Testament aus Talmud und Midrasch.* Munich: Beck.

Strasser, Susan. 1982. *Never done: A history of American housework.* New York: Pantheon.

Sylvain, Georges. 1901. *Cric Crac.* Paris: Ateliers Haïtiens.

Tamanoi, Yoshiro, A. Tsuchida, and T. Murota. 1980. The earth as an open steady system. Department of Economics draft. Okinawa International University.

Tarr, Joel A., and Francis Clay McMichael. 1977. Decisions about wastewater technology, 1850-1932. *Journal of the Water Resources Planning and Management Division* ASCE 103:47-61.

Teich, Mikulas. 1982. Circulation, transformation, conservation of matter and the balancing of the biological world in the eighteenth century. *AMBIX* 29:17-28.

Tellenbach, Hubert. 1956. Die Räumlichkeit der Melancholischen: Uber Veränderungen des Raumerlebens in der endogenen Melancholie. *Nervenarzt* 27:1.

Temkin, O. 1953. A historical analysis of the concept of infection. In *Studies in Intellectual History.* Baltimore: Johns Hopkins University Press. 123-847.

Thorndike, L. 1928. Sanitation, baths and street cleaning in the Middle Ages and the Renaissance. *Speculum* 3:192-203.

Thuillier, Guy. 1961. Pour une histoire de la lessive en Nivernais au XIXe siècle. *Annales ECS* 24:377-89.

Bibliography

———. 1968. Pour une histoire régionale de l'eau en Nivernais au XIXe siècle. *Revue d'Histoire Economique et Sociale* 46:232–53.

———. 1977. La lessive. In *Pour une histoire du quotidien en Nivernais*. Paris: Mouton. 127–37, 364–73.

Turner, John F. C. 1976. *Housing by people: Toward autonomy in building environments*. London: Marion Boyars.

Trumbull, H. Clay. 1896. *The threshold covenant*. New York: Scribners.

Ullmann, Manfred. 1970. Medizin im Islam. *Handbuch der Orientalistik*. Ergänzungsband 6.1. Leiden: E. J. Brill.

Van Liere, Eldon H. 1980. Solutions and dissolutions: The bather in nineteenth-century French painting, the image of the bather. *Arts Magazine* 54:104–14.

Varrentrapp. 1868. *Entwässerung der Stadte, Wert und Unwert des Wasserklosette*. Berlin. Cited in Schulz.

Verdier, Yvonne. 1979. *Façon de dire, façon de faire: la laveuse, la couturière*. Paris: Gallimard.

Vernant, Jean-Pierre. 1959. Aspects mythiques de la mémoire en Gréce. *Journal de Psychologie*. 1–29.

Vogue, A. de. 1971–1972. La règle de S. Benoit. *Sources Chrétiennes*. Vols. 181–86. Paris: Editions du Cerf.

Vuarnet, Jean Noël. 1980. *Extases féminines*. Paris.

Webster, Charles. 1976. *The great instauration: Science, medicine and reform, 1626–1660*. London: Holmes and Meier.

Wilson, T. P. 1978. *The Oxford dictionary of English proverbs*. Oxford: Clarendon Press.

Wolfson, Harry A. 1948. *Philo: Foundations of religious philosophy in Judaism, Christianity, and Islam*. 2 vols. Cambridge: Harvard University Press.

Wright, Gwendolin. 1979. *Moralism and the model home*. Chicago: University of Chicago Press.

Wright, Lawrence. 1960. *Clean and decent: The fascinating history of the bathroom and the WC*. Toronto: University of Toronto Press.

Zahan, Dominique. 1970. *Religion, spiritualité et pensée africaines*. Paris: Payot.

Zeraffa, M. 1975. Présence de la ville dans l'écriture du roman: Aspects psychologiques et formels. *Journal de psychologie normale et pathologique* 2:191–210.

Ziegler, Joseph. 1937. "Dulcedo Dei": Ein Beitrag zur Theologie der griechischen und lateinischen Bibel. *Alttestamentliche Abhandlungen* 13, pt. 2. Ed N. Nikel and A. Schulz. Münster: Aschendorf.

Zimmer, Heinrich. 1938. Die indische Weltmutter. In *Eranos Jahrbuch* 6. Zürich: Rhein Verlag, 175–220.

Zucker, F. 1928. Syneidesis-Consciencia. *Jenaer Akademische Reden* 6.

Zwernemann, Jürgen. 1968. *Die Erde in Vorstellungswelt und Kulturpratiken der sudanischen Völker*. Berlin: Reimer.